CLIMATE CHANGE

CLIMATE CHANGE
The Human Aspect

Editor

Krishan Saigal

KALPAZ

Climate Change : The Human Aspect

Rs. 750

ISBN: 978-81-7835-832-1

Published in 2010 in India by
Kalpaz Publications
C-30, Satyawati Nagar,
Delhi-110052
E-mail: kalpaz@hotmail.com
Phone : 9212729499

Lasser Type Setting by: Rudra Computer Graphics, Delhi
Printed at : Chawla Offset Press, Delhi

Contents

Preface

Anthropogenically induced climage change is likely to be a mega-catrostrophy affecting adversely all aspects of human life and activity. Since the underlying causes like carbon emissions exceeding carbon absorption, as well as the release of other green house gases in the atmosphere are all created by human activities, the solutions have also to be found only by human beings. I am therefore happy that the human aspect of climate change has been dealt with in detail in this timely book. I congratulate the Society for Indian Ocean Studies for bringing out this book.

We should lose no further time in initiating concerted measures at the global, national and local levels in fostering both mitigation and adaptation measures. To quote Paul Crutzen, "In order to be safe, we would have to reduce our carbon emissions by 70% by 2015. We are currently adding 3% more each year".

We cannot mortgage the future of our children and grandchildren by not responding to the challenge of climate change. The steps recommended in this book should therefore be widely read and practiced. I congratulate Dr. Krishan Saigal on his efforts to bring out this publication in a reader friendly manner.

M.S. Swaminathan

Acknowledgements

Dr. M.S. Swaminathan has very kindly given a preface to the book and his contribution is gratefully acknowledged.

The Society for Indian Ocean Studies has given permission for publishing the articles presented in the seminar on "Climate Change: South and Southeast Asia" which had been co-sponsored by Nehru Memorial Museum and Library.

Thanks are also due to Dr. Surinder Sud for permission to publish a chapter from his book "*The Changing Profile of Indian Agriculture*".

Grateful thanks are also to the Journal of Foreign Affairs for giving us permission to reprint the article **"The Other Climate Changers Why Black Carbon and Ozone Also Matter"** by Jessica Seddon Wallack and Veerabhadran Ramanathan.

List of Contributors

1. Nitin Desai, former U.N. Under Secretary General for Sustainable Development.
2. P.V. Rao, Director, Centre for Indian Ocean Studies, Osmania University, Hyderabad, India.
3. S. K. Chawla, Adviser, University of Petroleum and Energy Studies, New Delhi.
4. Malti Goel, former Adviser, Ministry of Science and Technology, Government of India.
5. Col. P.K. Gautam, Research Fellow, Institute of Defence Studies and Analysis, New Delhi, India.
6. Shyam Saran, former Foreign Secretary, Government of India.
7. S. Venkatesh, Joint Director & Head, Division of Epidemiology, National Institute of Communicable Diseases, Delhi.
8. Kakoli Banerjee, Department of Marine Science, University of Calcutta, Kolkata, India.
9. Avijit Gangopadhyaya, School of Marine Science and Technology, University of Massachusetts, Dartmouth, USA.
10. Issac Manuel R, M.S. Swaminathan Research Foundation, Chennai, India.

11. Juli and Vivek Cariappa, Organic Farmers, Halasur Village, H.D. Kote Taluk, Karnataka, India.
12. Sikka, former Head of India's Climate Research Programme, Department of Science & Technology, Delhi.
13. Surinder Sud, Agricultural Editor, Business Standard, Delhi, India.
14. Jessica Seddon Wallack, Director of the Center for Development Finance at the Institute for Financial Management and Research, Chennai, India.
15. Veerabhadran Ramachandran, Professor of Climate and Atmospheric Studies at Scripps Institute of Oceanography, San Diego, California, USA.

1

Introduction

Krishan Saigal

Global warming and the more acceptable term, climate change, has been on the international and national agenda for quite some time. There was a difference of opinion among scientists as to what had led to this state of affairs. There was intense debate in the 1980's as to whether global warming (the average global temperatures have risen by 0.7°C since the pre-industrial era) was due to natural causes or due to human activities. Scientific evidence was used by powerful lobbies, which stood for unrestrained growth and development to underpin the view that geological and other evidence suggested natural causes rather than human interventions. The green and ecological lobby challenged this assessment but the developed countries, especially the USA, supported the pro-development view and refused to be part of a global effort at restricting emissions of carbon dioxide into the atmosphere.

To get a scientific consensus on the issue, the UN set up the Inter-Governmental Panel on Climate Change (IPCC) with more than 2500 scientists from both developed and developing countries to go into the scientific evidence and give a consensus assessment on:

(a) whether global warming was taking place;

(b) whether it was due to human intervention, and

(c) to make recommendations as to what could be done to mitigate and/or adapt to the likely impact of climate change.

The IPCC released its Fourth Assessment Report in 2007. The Report deals with the probabilities and risks involved in understanding and relating to the consequences of climate change. The Report comes to the view that there is global warming, that it is due to anthropogenic (human) activities and that if no action is taken in the next couple of decades, temperatures could go up by as much as 5°C by 2100. To put this in perspective, a 5°C rise is the same that melted away a 1km ice sheet in North America and Europe at the end of the last ice age. The IPCC further goes on to stress that the threshold for dangerous climate change is an increase of around 2°C. This threshold broadly defines the point at which rapid reversals in human development and a drift towards irrevocable ecological damage would become very difficult to avoid.

The early warning signs are already visible. We are witnessing today what is happening to the world's poor in the developing countries. Increased exposure to drought, to more intense storms and environmental stress is holding back the efforts of the poor to build a better life for themselves and their children. As sea levels rise and tropical storms (cyclones and tsunamis) gain in intensity and frequency, millions face displacement. Dry land inhabitants, who are some of the most vulnerable, have to cope with more and sustained droughts with consequences on food security, malnutrition and spread of diseases. As glaciers recede, water supplies are being put at risk. The gender discrimination is also widening, as women have to travel long distances to collect water and fuel.

The recommendations of the IPCC deal with a number of inter-related issues. There is, firstly, the need for putting a cap on greenhouse gas emissions. This requires deeper emission reductions by the developed countries. Second, there needs to be serious engagement with the developing countries, as well as incentives for them to limit their emissions while safeguarding economic growth and efforts to eradicate poverty. It also emphasizes the need to strengthen the framework for international cooperation. It is necessary to enhance access to modern energy services and reduce dependence on biomass, i.e, coal, the primary source of energy for the poor. Significant strengthening of international cooperation on coal with greater development and deployment of the Integrated Gasification Combined Cycle technology is necessary. This would require financial incentives on a sustained and international basis.

Based on the recognition of the fact that in spite of any mitigation efforts made in the near future, due to the time lag between action and its result on climate change, temperatures are likely to rise till 2050. So it is necessary to include adaptation as a key element of international efforts. It is necessary to enhance the capacity of developing countries to assess climate change risks and their ability to integrate adaptation into all aspects of national planning. Poverty reduction programmes have to go hand in hand with building the resilience of vulnerable sections, and empowering them, through information, education, health and social protection measures, like poverty eradication and employment generation programmes. There is also need for a major push to generate new technologies to combat climate change, to make existing renewable technologies economically viable, and to promote a rapid diffusion of technology.

The IPCC Assessment takes a global look at the climate change scenario. All its recommendations are meant for the governments of states. Their recommendations were discussed at Bali, and later on will be discussed in Tokyo, to get a global consensus on the question of emissions, technology, international cooperation, etc. on the laid down principle "that the polluter pays" and that "there should be common but differential responsibilities" of the different states. In these global conferences the stand of India, as indeed of other developing countries, has been the principled one of asking the presently developed states, who have left the largest carbon footprints in the atmosphere, to take on the responsibility of making deep cuts in their green house emissions. The developing countries would also like the developed world to take on the task of developing energy-efficient technologies and of transferring them to the developing states.

The findings of the IPCC were examined further by a United Nations Development Agency. It examined the impacts of the likely changes in climate on human development (the Human Development Report or HDR, 2007-2008). In the meantime, as the HDR points out, the impacts of climate change are already being felt by the developing countries and especially the poor and vulnerable sections. Though climate change is a global phenomenon, the human development impacts of climate change cannot automatically be inferred from global scenarios. People (and countries) vary in their resilience and capacity to manage the incremental risks associated with climate change. They similarly vary in their capacity to adapt.

With the global rise in temperature, local rainfall patterns are changing, ecological zones are shifting, the seas are warming and ice caps are melting. Developing countries like India and the rest of South and Southeast Asia are at a double disadvantage; they are located in the tropical zone that stands to experience some of the most severe early impacts from climate change; and agriculture — the sector most severely affected — plays a far greater social and economic role. Above all the poor in this area are characterized by high levels of poverty, malnutrition and disadvantage in the provision of health facilities.

Climate change, though global in nature, has effects which are local. Physical impacts will be determined by geography and micro level interactions between global warming and existing weather patterns. For example, in India the timing, intensity and pattern of monsoons are likely to be affected leading to drought in certain area with severe flooding in others. This is already happening leading to acute pressure and stress on the physical, social and economic infrastructure. Human development impacts will vary from region to region but at least five specific risk multipliers in South and Southeast Asia can be identified.

First, there is a likelihood of decreased agriculture productivity due to drought and rainfall variations. This will impact on the production of food crops, lead to losses of income as well as severe malnutrition especially of women and children.

Secondly, there is going to be a heightening of water insecurity. Glacial melt in the Himalayas will compound the problems of water ecology in northern India leading initially to heavy flooding followed by reduced water flow for the irrigation systems.

Thirdly, there is likely to be coastal flooding coupled with extreme weather events. These could affect West Bengal, Bangladesh and the Mekong Delta region as well as other areas subject to cyclones and tsunamis like Indonesia, Malaysia and the Philippines.

Fourthly, there could be the collapse of various ecosystems like mangroves, coral reefs, etc. Coral reef systems already in decline would suffer extensive 'bleaching' leading to the transformation of marine ecologies, with large losses of biodiversity and ecosystem services. This would affect the millions dependent on fish for their nutrition and livelihoods.

Lastly, there are the increased health risks. There would be an increased possibility of contracting malaria, dengue, cholera, dysentery and other water-borne diseases.

The HDR points and that these five factors should not be taken in isolation. They interact with each other, and with existing problems like lack of social, administrative and health infrastructure to produce a vicious downward spiral. The poor, women and children would be the main victims, especially those living in the rural areas.

To examine the likely changes in climatic pattern on the populations living in the countries of South and Southeast Asia, the Society for Indian Ocean Studies and the Nehru Memorial Museum and Library brought together in a seminar in February, 2008 a multi-disciplinary group of scholars, researchers and practitioners to go into the issues involved especially as they affected the poor and vulnerable sections. Apart from the papers presented at the seminar, papers have been commissioned to fill gaps. The two recent reports of the Intergovernmental Panel on Climate Change (IPCC) and the Human Development Report of UNDP (HDR) were taken as the starting point for delineating the regional issues involved both in the realm of state policy and action as well as what could be done by the civil society. Stress was laid on adaptation in addition to the technological and scientific issues involved. The papers present at the seminar have since been revised to bring them up to date. A number of fresh papers have also been commissioned to cover as wide a field as possible.

The first paper by Nitin Desai, 'Impact on South and Southeast Asia,' looks at the likely impacts on South Asia which he identifies as the most vulnerable on account of the possible collapse of the summer monsoon. Besides, the retreat of the Himalayan glaciers would lead to major rivers like Brahmaputra, Ganges, Indus, Yamuna and the rivers of Punjab becoming seasonal streams. Sea level rise would very adversely affect Bangladesh and parts of West Bengal. He opines that the impact on Southeast Asia would be similar with the retreat of the Tibetan glaciers instead of the Himalayan ones.

Shri Desai goes on to say that climate change cannot be tackled by any nation on its own. There is need for global as well as regional agreements so that cooperative actions can be taken. In his opinion, elements of such an agreement could be the establishment of an acceptable limit for rise in global temperatures. He thinks this should be 2 degrees centigrade instead of 3 degrees as is being generally discussed. This has to be underpinned by the time profile of restraints by each major player and not left hanging in mid-air as it were. While appreciating the stand of India and the developing countries on the need for looking at cumulative emissions since the start of industrialization so that the developing countries are allowed to have the same footprint as the developed countries, he is of the opinion that this stand would not stand for long. To take the position of India, its gross national product would be 8 to 10 times what it is today by 2050. Today's India adds only about 2 tonnes per capita of Green House Gases (GHGs). A figure of 8 to 10 times that amount would obviously not be sustainable. So he thinks that keeping considerations of equity in mind, Indian and other developing countries have to accept some obligations on limiting GHGs.

Besides, the poor in these countries are already under stress. It is necessary to adapt to climate change as is already being done by the poor all over the region. Failure by these countries to act would very adversely impact the poor especially women and children. Women would have to go long distances to collect fuel and water while malnourished children would be vulnerable to various diseases. Failure to act by the countries of South and Southeast Asia could lead to disaster especially for the vulnerable sections.

The paper by P.V. Rao, "ASEANS approach to Climate Change" examines the efforts being made on a regional basis by the ASEAN countries to meet the new challenges posed by rapid economic growth, growing urbanization and rising energy consumption. Concerned at the rapid deterioration of the environment, the ten ASEAN nations at a meeting held in Singapore in September 2007 decided to take the following adaptation and mitigation strategic to minimize the impact of climate change:

- to encourage the deployment of clean technology
- to commission studies both national and regional for assessing the impact of climate change on the environment.
- to regionally cooperate in developing carbon mitigation and cleaner fuel technologies.
- to cooperate for the development and use of civilian nuclear power.
- to cooperate and reduce deforestation and forest degradation while simultaneously promoting afforestation and reforestation.

Rao argues that the Singapore declaration was the result of the environmental degradation arising and of rapid economic growth as well as the on going debates and global concern about climate change. Only the future will clarify the extent to which ASEAN countries would reconcile their regional concerns with subscription to international conventions and agreements.

In his paper 'Impact on the Indian Ocean Region', S.K. Chawla considers the links of energy, which, according to him, is the main driver of the economy, with climate change. Since fast economic growth in the underdeveloped countries of the Indian Ocean Region is absolutely necessary, steps are necessary to delink, to some extent, climate change from the adverse impacts of growing energy requirements. There is no simple remedy but energy saving measures (for example, through increased efficiency in energy utilization) have to be complemented through the use of carbon dioxide free technologies for energy production. There is also the need to increasingly replace the use of fossil fuels by the large-scale deployment of the new renewables like solar, wind, geothermal, etc. There is also the need for increased worldwide use of nuclear energy. In this connection because of the large investments involved, Chawla recommends that the developed countries should take the lead in this regard.

There is also the need to discourage the vulgar display of wealth by measures like rationalization of the transport sector and development of infrastructure on an equitable basis. A drastic change in life-styles, and recycling of scarce resources like water harvesting and sewage management is necessary. This would require parallel actions on the social front to complement the new technologies that need to be developed. Action is needed by civil society so that awareness is created on energy, environment and health. Chawla calls on welfare societies, religious and spiritual organizations, schools and colleges to take the lead in this regard so that an appropriate and environmentally sensitive culture is created.

Malti Goel in her paper, 'Technologies for Carbon Dioxide Sequestration', deals with methods for removing carbon dioxide from the atmosphere. Carbon dioxide sequestration (carbon capture and storage) is a scientific and technological approach towards reducing net concentrations of carbon dioxide in the atmosphere. Essentially carbon capture and storage has three main components: capture of carbon dioxide from large point sources; fixing it or transporting it to a possible location where it can safely stored; and finally processing it and fixing it.

The paper gives a summary of technologies available for carbon dioxide capture and storage. Most of these are being researched and in many cases their cost-effectiveness is in doubt. These technologies include capture from industrial and power plants, capture from natural gas whether post-combustion or pre-combustion, and capture from the atmosphere. The captured gas is liquefied for ease of transportation and thereafter transported to possible geological and terrestrial storage sites. Research is also on for storing the compressed gas at great depths in the oceans. But the consequences of this storage on marine organisms is still to be studied.

For storage on land, oil and gas on-shore and offshore empty fields, un-mineable coal seams, aquifers and terrestrial forests are considered to be suitable. Possibilities also exist in bio-fixation through microbial and micro-algae based processes but these are still in the process of research. Terrestrial and agro-forestry model studies are under way since this is a natural and cost-effective method of fixing carbon dioxide while giving increased production of timber and crops as a by-product.

More research and development efforts are necessary before the technologies can be made cost-effective and environmentally acceptable. The article suggests collaborative research between the countries of South and Southeast Asia as the region has many similarities and needs.

The next two articles deal, among other things, with the geopolitical aspect especially the economic and other impacts on the populations in developed and developing countries. P.K.

Gautam in the article, 'climate change, environmental degradation and societal challenges', after dealing with the common ecological challenges arising out of climate change states that the adaptation capacity of the countries of South and Southeast Asia is limited due to poverty, and constraints on capacity to cope with disasters, etc., a low economic base. This region is also among the worst affected areas due to its location in the vulnerable tropical zone. Sea level rise, for example, could lead to submergence of the Maldives while sea intrusion into Bangladesh, parts of India and Sri Lanka could lead to the problem of migration leading to social stresses, and even geopolitical stress if the migration is from one country to another.

Increase in frequencies of floods and drought due to retreat of glaciers could lead to decline in productivity in the agriculture sector leading to shortage of food, fodder and fuel. There would also be an adverse impact on infrastructure, especially coastal infrastructure, due to cyclones, tsunamis and sea level rise. The impact on human health due to rise in vector borne diseases would be especially severe on children who are likely to be malnourished due to shortage of food.

The author suggests initiation of dialogue between the countries included in the South Asian Association of Regional Cooperation (SAARC) on climate related adverse impact especially on the shared water systems as well as on measures to deal with disasters. There is also need to treat climate change and environmental challenges as an opportunity to focus on the necessary social and behavioural changes. There is the Gandhian model of frugality and austerity which can be taken as a reference point so that demand needs can be controlled instead of letting the market forces and the consumerist ethic have uncontrolled sway. Civil society as represented by voluntary agencies and intellectuals could play a decisive role in this regard.

Shyam Saran in his article 'Climate Change – Impact on Economic Development', deals with the costs to be incurred by countries like India on adaptation measures (estimated to be currently 2.5% of the national economic product) as well as on changing the patterns of activity in agriculture, industry and the sciences. This would adversely affect economic and social development and the prospects of eradicating poverty within a generation. There would in addition be costs to be incurred if concerns over climate change compel a significant and accelerated shift in the energy mix away from the current reliance on fossil fuels to non-fossil fuels.

Saran argues against the developing countries taking on of commitments to reduce emissions of green house gases. He adduces data to show that leaving aside the cumulative accumulation of green house gases in the atmosphere even today the developed countries are adding per capita emissions nearly 30-40 times that which is being emitted by, say, India! Therefore, the negotiating strategy of the developing countries in the climate negotiations should be not to accept any inequitable regime, as that would place a serious constraint of their growth processes.

The article also quotes an expert report to argue that for economic and supply reasons, a shift from renewable to non-renewable sources of energy is necessary. And in this case a by-product would be its impact on climate change. But this should be done autonomously and not as a result of pressure from the developed world.

Dr. S. Venkatesh deals with the impact of climate change on human health. He envisages impact on human health as increases in the following:

- deaths due to sudden heat waves, floods and droughts
- vector-borne diseases such as dengue, malaria and chikungunya

- water-borne diseases especially diarrhoeal diseases
- malnutrition and its associated effects on child growth and development
- morbidity and mortality related to cardio-respiratory ailments.

For example, it has been estimated that in India the proportion of the population exposed to dengue would rise from 30% to 60%. Similarly, uncontrolled floods and contamination would lead to diarrhoeal diseases while migration of people due to floods and drought would make them vulnerable to various communicable diseases.

Some steps have been taken by Bhutan, Thailand and India for early warning systems based on forecasts. They have established medical response teams, strengthened programmes aimed at controlling vector-borne and water-borne diseases and have empowered local communities through advocacy and awareness – increasing measures. But more needs to be done and the health programme needs to be integrated with other programmes aimed at adapting to the changes brought about by climate change.

The last three articles deal with some local areas in India and how people have reacted to and adapted to the impact of climate change. In their paper on 'Impact on the Gangetic delta', Banerjee, Gangopadhyay and Mitra look at the impact of climate change on an estuarine system, namely, the Gangetic delta region. They approach the problem through an integrated approach as the area is the final receptacle of all waters from the upstream zone, river discharges and run-off. To get a realistic picture of the climate change they collected data on surface water temperature, salinity and dissolved oxygen after compartmentalizing the deltaic complex into east and west. While the western rivers get their water from the Himalayan glaciers, the eastern rivers are tide-fed.

The temperature data showed rise in both the western and eastern parts. Lowered salinity and increasing dissolved oxygen in the western part pointed to increased fresh water supply due to the melting of the Himalayan glaciers. In the eastern part global warming was apparent through increased tidal amplitude, increased salinity and lower dissolved oxygen due to intrusion of saline water from the Bay of Bengal.

These different footprints of climate change are posing serious problems in the livelihood and biodiversity patterns in the area. This is leading to migration to nearby industrial complexes and urban areas. Cyclonic activity which is endemic to the area would lead to large scale displacement of human and animal population.

Isaac Manuel in the paper, 'Enhancing adaptive capacity to climate risks' gives details of a project started by the M.S. Swaminathan Foundation to study the vulnerability profiles and farmer decision processes and adaptation choices under adverse conditions in their farming practices. Experiences, lessons, mechanisms and techniques to enhance adaptation at the local levels are being identified. The idea of adaptation is to reduce the adverse consequences of climate change, or to take advantage of any opportunities it may present.

Besides the coastal zone, arid and semi-arid are considered to be most vulnerable to climate change. Rajasthan and parts of Andhra Pradesh, which are in the arid and semi-arid zone, were selected for the operation of the project. The selected areas experienced an annual rainfall between 100 and 400 mm in the arid zone and 400 mm to 800 mm in the semi-arid zone with a variation coefficient of 40 to 70%. Low and erratic rainfall coupled with extreme temperatures and intense solar radiation results in frequent crop failures and considerably affects the agricultural economy.

The project has three objectives: to build community level capacities in the agriculture, water and energy sector; to optimize the service delivery system; to promote general awareness on climate issues. To achieve these objectives the programme combines four different components that help to define the way of implementing activities: research and analysis; natural resource management; capacity building and action learning; and local empowerment.

The research and analysis component covers the development of a local knowledge base on the effects of climatic hazards, on the availability of water and energy resources and on agriculture production systems. It aims at understanding the strategies and techniques that local communities have developed to cope with these climatic hazards.

The resource management component covers interventions that aims at developing management practices on a pilot scale. It includes the participatory development of techniques and technologies as well as social agreements for managing water, agricultural production, and distribution at community level. Improving management skills, validating existing techniques, transferring best practices and producing management guidelines and manuals for extension workers is a part of this component.

The component of capacity building and action learning aims at improving awareness and in-depth understanding of vulnerabilities and the adaptation. The local empowerment component, in addition to what has already been described, aims at improving knowledge and negotiating skills. To do this the programe includes activities to improve the information flow, the functioning and creation of a platform for policy advocacy, and the design and use of information systems. All the components are complementary and are designed to promote holistic understanding.

The programme hopes to make an important contribution to 'Adaptive Science' and to demonstrate and implement adaptation and coping measures at the local level.

Juli and Vivek Cariappa in their paper, 'Sustainable solution in organic farming', point out that while it took thousands of years to change-over from hunter-gatherers to nomads to peasant-farmers, the change-over from peasants to farm-managers took only fifty years. The system of cultivation has changed. This has meant over-using precious non-renewable fossil fuels, over-exploiting the topsoil both physically, in terms of in-appropriate ploughing, as well as chemically, in terms of disturbing the balance and availability of atomic elements. Mono cropping, massive and wasteful use of petroleum based agro-chemicals which require heavy irrigation, soil compaction due to use of heavy machinery and the lack of biomass returned to soil to replace lost humus, has meant a huge loss of carbon directly and indirectly thus contributing significantly to global warming.

The effect of the different pesticides that accompany this model has deeply affected the biological checks and balances required for a resilient agriculture. The authors point out that despite an 80% increase in the world chemical fertilizer use since 1950, there has been only a 53% increase in yield accompanied by a sharp drop in nutritive quality of food, e.g., a 17% drop in the content of wheat protein has been recorded in the U.S.A.

Organic farming, in contrast, is based on renewing the living soil and re-creating through its inherent need for biomass, an oxygen rich environment which incorporates carbon rather than releasing it, while absorbing moisture for slow release. By its very nature the farming system imitates the forest ecology from which we have all evolved.

The authors regard organic farming as what is required to meet the challenge of climate change. In their view, organic farming has to be seen only as a system of food production but also in terms of its wider effects of increasing the carbon content of the soil, increasing its water holding capacity and its ability to sustain a diversity of plants and living organisms to balance a closed circuit of energy on Earth. Organic farming also optimizes the energy ratio, so that the energy expended in the production system leads to much longer output through photosynthesis and the conservation of water.

The authors have been doing organic agriculture for the last thirty years and have many suggestions to offer on pest control, mix of crops, use of tree cover in the tropics and so on. They also use heavily on green manuring and cropping systems which utilize different yet mutually beneficial types of crops including horticultural and plantation crops including inter-cropping. The authors conclude by stating that organic farming offers practical and plausible solutions to not only slow down the rate of climate change but adequately feed the world's population.

The Monsoon is of critical importance to all the countries of South and Southeast Asia as most farmers depend on the rains. Dr. Sikka has given a technical account of how the Monsoon interacted with global developments whether the melting of the Arctic ice or developments in the Pacific Ocean.

Dr. Sud has extensively brought out the impact of climate change on agriculture in India. He has also highlighted the steps being taken by research scientists both as to how to forecast the Monsoon as well as the crops and practices necessary to adapt to global warming. He points out that the impact of global warming on the production of cereals is particularly worrisome. He stresses the need to shift to agro-forestry, enhancing the efficiency of nitrogenous fertilizers and to have enhanced capacity for agriculture-oriented seasonal forests so as to have more effective measures for timely delivery of suitable crops etc.

Wallack and Ramanathan have in their article drawn attention to the low-risk, cost-effective, and high-reward option: reducing emissions of light-absorbing carbon particles (known as "black carbon") and of the gases that form ozone. Together, these pollutants' warming effect is around 40-70 percent of that of carbon dioxide. Limiting their presence in the atmosphere is an easier, cheaper, and more politically feasible proposition than the most popular proposals for slowing climate change — and it would have a more immediate effect.

Three appendices giving relevant extracts from (i) the fourth IPCC assessment on Climate Change and (ii) the human development report 2007-08 are being attached as well as a copy of the UN Framework Convention on Climate Change as these form the background in which all countries have to act.

2

Climate Change: South and South-east Asia

Nitin Desai

We would not be here if there was no greenhouse effect. If we did not have an atmosphere around it, the Earth would have a temperature of around minus 18 degree centigrade. In practice, its average temperature is plus 14 degree centigrade. So there is a greenhouse effect of 32 degrees anyhow. We have also had changes in this 32 degrees warming over time. We had Ice Ages and we also had the end of Ice Ages. The real issue now is the pace of change and the reason for the change. Previous changes may have taken place over long periods because of the natural factors, such as changes in solar activity, volcanic activity, changes in the tilt of the axis of the earth, etc. What we are talking of today are changes which are essentially induced by human intervention. Most scientists are convinced that this is

---

* This is the edited of a transcript of a talk delivered at a meeting of the Society for Indian Ocean Studies held at the Nehru Memorial Museum and Library on 16 February 2008.

Chart 1 : Direct Observations of Recent Climate Change

Global mean temperature

Global average sea level

Northern hemisphere Snow cover

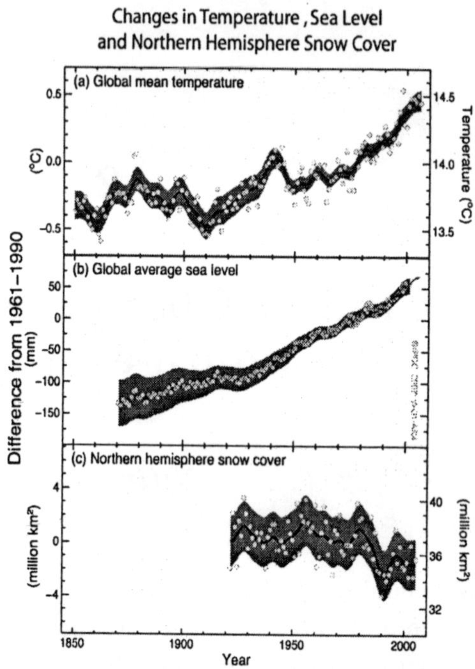

Source : IPCC WG I.

a reality, that the global warming we are experiencing is due to human induced factors rather than some natural cycle which may well reverse itself. There are a few eccentrics who continue to believe the latter, but they are a small minority.

The direction of climate change from around 1860 onwards can be seen in terms of certain key parameters. Chart 1 shows the global mean temperature, the average sea-level and, what is very important, the snow cover in the Northern hemisphere. You can see the pace of change and the acceleration of change in more recent times in this chart.

Chart 2 shows the record of temperature change. The red line shows the trend change since 1920. The much more steep

Chart 2 : Global Mean Temperature

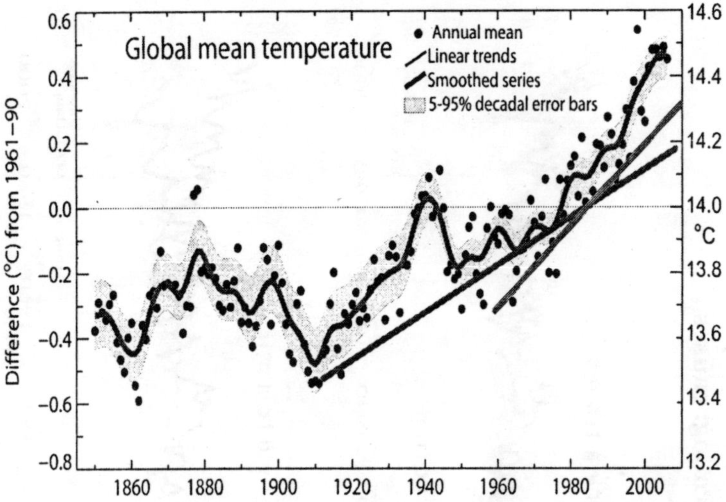

green line shows the trend change since 1960. This shows why scientists believe that the pace of global warming is accelerating.

Chart 3 compares the actual temperature change with what could be expected from natural causes. The blue line is what you would have expected from solar and volcanic activities. As you can see, the actual change is very much higher. This demonstrates very sharply and clearly that there cannot be any dispute what we are seeing is due to human intervention. It is important to assert this as there are still some people who feel that we have to wait and watch because this may be a natural cycle which will reverse itself. The warming that we are experiencing has to be attributed in large measure to the greenhouse gases that human activities have emitted into the atmosphere since the start of the industrial revolution. These are carbon-dioxide which comes mainly from the use of fossil fuels, methane which comes partly from energy production and partly also from agriculture and nitrogen oxide mainly from fertiliser use in agriculture.

Chart 3. Global Warming: causes

- Observed changes consistent with

☑ expected responses to forcings

☒ inconsistent with alternative explanations

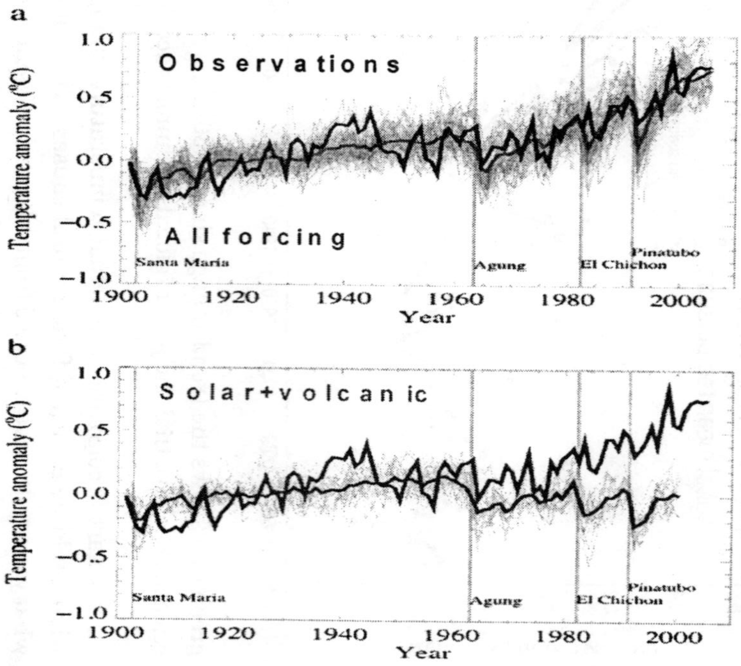

Chart 4 : Projections of Future Changes in Climate

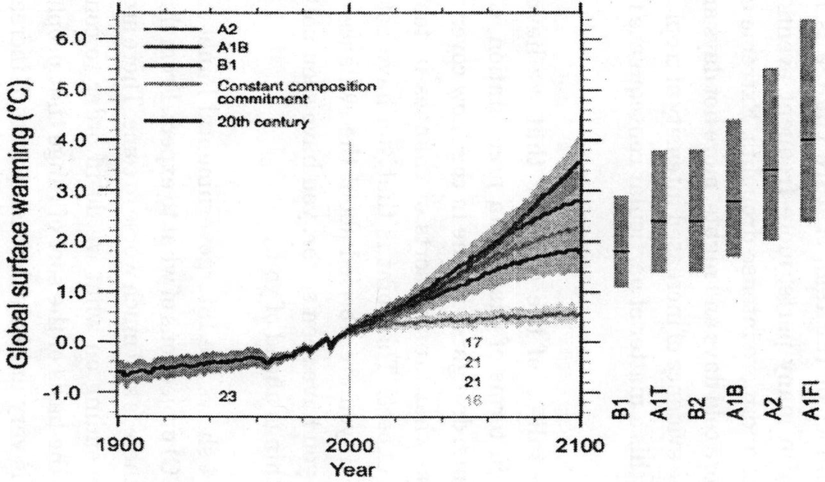

Best estimate for low scenario (B1) is 1.8°C (likely range is 1.1°C to 2.9°C), and for high senario (A1F1) is 4.0°C (likely range is 2.4°C to 6.4°C).

Source : IPCC WGI for AR4

The impact of the temperature change is increased precipitation in many parts, more frequent events of heavy precipitation, and more intense droughts. Extreme events are also up — more cold days and nights, more hot days and nights. There is some evidence of increased intensity of tropical cyclone activity and this a matter of particular consequence for coastal regions.

I have talked of the changes that we have already experienced in terms of changes in precipitation, changes in extreme events, droughts, decline in the snow cover and so on. Beyond these what are the sorts of changes in temperature that we can expect? The point is that you have not seen the worst as yet. If you are worried about the evidence of climate change that you have seen so far, you have seen nothing. You have worse things ahead of you.

Chart 4 shows the Intergovernmental Panel on Climate Change (IPCC) projections of what to expect. From this graph it is clear that there is still much worse to come. There are different scenarios depending on what we do in order to contain that damage. On the basis of the sort of range that people are now talking of it is very unlikely that temperature increase can be contained below 1.5 degree centigrade. It is possible that it may go up beyond 5 degrees centigrade.

If we do not act so as to contain the atmospheric concentration of Green House Gases (GHGs) temperature increase could go to a point where we run the risk of runaway warming with catastrophic consequences. But even short of such runaway warming we run the risk of crossing some crucial tipping points or extreme events that will affect our environment very unfavourably. Some of these are as follows with the time period over which the event would take place, if it happens, indicated in brackets:

- Melting of Arctic sea-ice (about 10 years)
- Decay of the Greenland ice sheet (about 300 years)
- Collapse of the West Antarctic ice sheet (about 300 years)
- Collapse of the Atlantic thermohaline circulation (about 100 years)
- Increase in the El Nino Southern Oscillation (about 100 years)
- Collapse of the Indian summer monsoon (about 1 year)

The important point to recognize is that we do not know whether these extreme events will take place or not. Right now, Arctic ice melt has begun but none of these other extreme events are projected to happen. But there is a risk that they could happen, if we do not do anything.

Of the extreme events, I would really like to draw your attention to the last one viz. collapse of the Indian summer monsoon. That is not projected but it could happen if we do not address climate change and, if it does, it would happen in one year posing huge problems of adjustment for us. Some other tipping points stretch out over a hundred years and more, so that more time is available to adjust. The collapse of the Atlantic Thermohaline circulation is the famous Gulf Stream and if that were to collapse Europe would become a cold desert. If you were to get the collapse of Antarctic ice-sheet or the decay of the Greenland ice-sheet we are looking at sea-level rise which is not just one meter which is being projected now. It will go up to six to seven meters. Any ocean scientist would tell you that six to seven meter rise is real disaster.

The impact on India, which will of particular interest to this audience is presented in the table below:

You can see that some regions should expect a significant rise in average temperature, and a significant rise in precipitation. You will see floods mainly in the Eastern part of

REGION	*TEMPERATURE Annual Average ℃*		*RAINFALL Average mm / day*	
	1961-90 Present	*2070-99 Projection*	*1961-90 Present*	*2070-99 Projection*
North-East	20.54	24.54	3.51	4.23
North-West	23.55	27.52	1.58	1.97
South-East	26.76	30.06	3.05	3.42
South-West	26.23	29.32	3.04	3.47

Source: Cline William R., Global Warming and Agriculture: Impact Estimates by Country, Peterson Institute for International Economics, Washington DC July 2007.

Chart 5 : Retreat of Gangotri Glacier

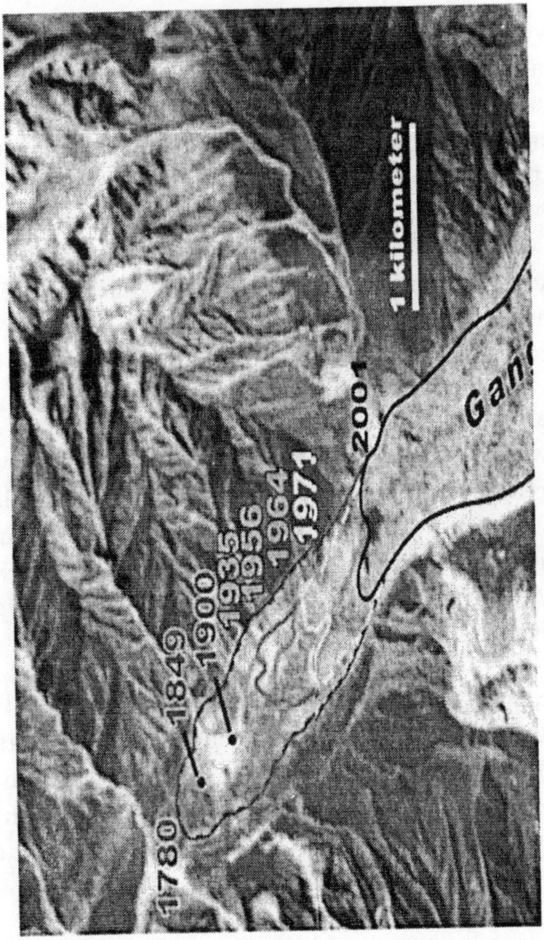

Rate of retreat for three decades three times the rate for previous 200 years.

Source : An Overview of Glaciers, WWF Nepal Program, March, 2005.

the country, and droughts more in the Western part of the country. You are going to see a situation where the River basin likes Banas, Sabarmati, Tapti, Mahi face shortages, whereas rivers like Brahmani, Godavari, Mahanadi going towards the Bay of Bengal would actually have floods.

A very crucial problem for us is the impact on the Himalayan glaciers. As many of you have read, the impact on the Himalayan glaciers is crucially important because the glaciers holdback water and release it gradually through the year. So the melting of the glacier, and the shrinking of the glaciers has two effects. One, it changes the time cycle of the water flow and two it reduces the overall flow. More water in spring and less in summer. You are looking at the possibility of the Yamuna in Delhi having only 30 to 50% of the water it has now in the dry season.

Chart 5 is a picture of the Gangotri basin. You can see where it was in 1971, and where it was in 2008, and how much it has retreated. It has been retreating for a long time, but rate of retreat has gone up. Most of us are old enough to remember the photograph of Tensing Norgay and the photo of the Kumbhu glacier before it started to decline. The foot of the Glacier is now five Kms. away. That is in our lifetime.

Chart 6 shows the impact of a 1.5 meter sea-level rise on Bangladesh. The sea would come much closer to Dacca. It shows the area affected if you have a 1.5 meter sea rise. Some 15 to 16% of the land area and population would be affected, and these people have to shift from there to somewhere else. The place where they are going to move is India.

The broader picture of Asian impact is very similar to the impact on South Asia with the possible exception of North-Asia, which shares with many other Northern regions the possibility that climate change impact may not be as adverse for it as for the rest of Asia.

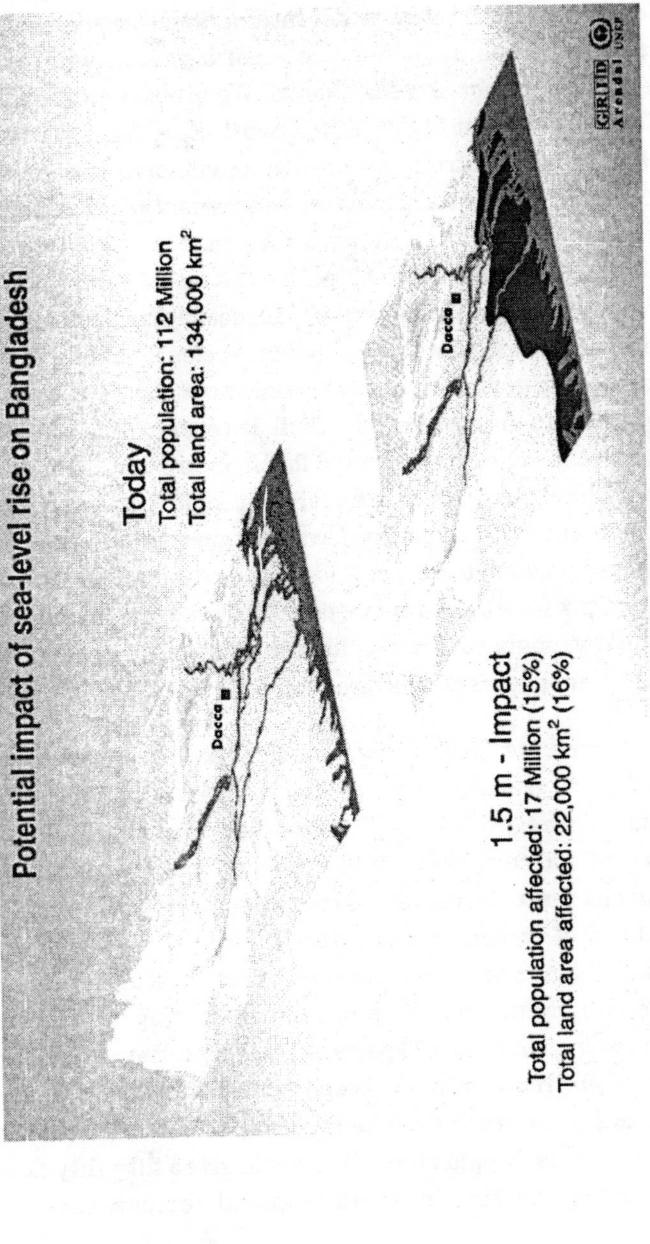

Source : UNEP/GRID Geneva; University of Dacca; JRO Munich; The World Bank; World Resources Institute, Washington D.C.

Climate change is not something that any nation can tackle on its own. We need a global agreement for cooperative action. What are the elements of International agreements that we need to focus on? First, we must agree on what is the acceptable limit for climate change. We are a very vulnerable region as is the rest of South and South-East Asia. We should have therefore, a strong interest in an effective international agreement on climate change. I keep stressing this because there is a persistent tendency on the part of our negotiators sometime to be happy if you have a weak agreement. This is an extremely shortsighted approach. I am not saying that we should take all the responsibility. But we must clearly assert that we need a strong global agreement. Remember this, the Northern countries do not have as much interest. The cost and consequences of climate change for them are not as drastic as they are likely to be for us, and for the people in this region. We cannot wait for more research, wait for more information, wait for more knowledge etc. I would therefore, point out that first we must ask for a tight limit on potential temperature increases. I believe strongly that it should be around 2 degree centigrade rather than 3 degrees centigrade.

Second we must agree on a time profile of restraint on emissions of greenhouse gases which consist of carbon-dioxide, methane, nitrogen oxide and a few others. But mainly, it is the restraint on carbon-dioxide emission which is the most important. Since the industrial revolution we have released 2000 giga tons, of carbon as measured in terms of carbon-dioxide. About 1200 giga tons was absorbed by eco-system globally, 800 giga tons accumulated in the atmosphere leading to an increase in concentration from 280 parts per million carbon to 380 parts per million. If we want to keep global temperature increase down to 2° centigrade then we can go up to about 450 parts per million, that is 70 ppm more. This would give a fifty-fifty chance of limiting the rise in average global temperature to 2°

centigrade. This goal means that ecological space available to us and everybody else in the world is now limited and the challenge is to agree on how it should be fairly shared.

The argument from India has been that we should be looking at cumulative emission. After all climate change is not just a function of how much carbon is emitted today and in the future. It is a function of what is the stock which exists in the atmosphere, and liability should be defined in terms of how much each country has contributed to this problem over the last two-hundred years. What is important is that neither India nor China can seriously argue that we are not going to do anything till our per capita cumulative emission reaches the industrial countries' level. The combined population of India and China is ten times the population of the industrial world. And since they already have emitted close to 1600 billions tons of carbon are we going to argue that we will not do anything till we emit 16000 millions tons carbon? Because in that case, there won't be any thing left. That is not a maintainable proposition even from a narrow nationalist point of view. The past has to be taken into account in terms of the liability for the cost of mitigation and adaptation. But the sharing of available space will have to rest on more workable principles of equity. Prime Minister Dr. Manmohan Singh has accepted this point when he assured the world that India will ensure that its per capita emission always remains below the average of developed countries.

This commitment is a constraint on us and we cannot pursue business as usual. Western Europe's emission is roughly around 8 tons per capita. They are now committed to bringing it down and have indicated a goal of 80% reduction by 2050. This would bring their emissions down to 2 tons per capita. This would leave us a little headroom for expansion, because we are at a little over 1 ton of emission per capita, keeping in mind our announced goal that we will always remain below the average for the developed countries. Note also that China is close to 4 tons per capita. So some people argue: hide behind China, they are so much worse than us and that any deal they negotiate is bound to be favorable for us also. But that is just a facetious argument.

There is growing support for the idea of contract and converge which is that at some defined point in time, we must all move towards a relatively similar per capita emission, and that level should be consistent with the temperature increase which should not exceed some agreed level, say 2 degrees centigrade.

India has tended to argue that we are not going to take any obligations. But the pressure on India will go up. We are at 1 ton per capita today, but by 2050 our GDP will be about 8 to 10 times higher. Carbon emission would reach the 8 tons per capita level if our GHG emissions increase at the same rate. That is not going to be accepted. The carbon constraint emerging from global negotiation will start hitting India much sooner than you think. In any case, a very important point we need to recognize is that the peak in global emission has to come by around 2015 plus minus if we want to contain temperature increase to 2 degrees i.e. not 20 years from now, not 30 years from now, but just about 5-7 years from now.

The further point I would like to stress is the importance of adaptation. Even with 2°C, increase, there will be significant impacts which we need to account in our development. We are not doing that. Our water resource projects are being developed on the basis of historical data. Our port projects which will last 40 to 50 years, are being developed on the basis of historical information on tides. Basically, the one thing you can say is that water projects or the sea regimes that ports face are going to be different, and we are not taking that into account when we are designing projects. A part of the explanation is that we do not have enough detailed, knowledge. We of course need better information and adaptation capacity. We also need to spend much more on agricultural research, on water research and other areas. We will also have to spend on coastal defences in many areas. The amounts involved could be quite substantial. At the global level increasingly the dialogue is how the cost of this adaptation is going to be shared. One of the successes Indian diplomacy has had is getting that entered into the climate convention.

We do not have to take on unfair commitments which will constrain our potential developmental task. But we have to recognize that we are going to live in a carbon constrained world, that we cannot necessarily follow the energy path which the industrial countries followed over two hundred years. That is not sustainable, and the world would not tolerate it for long.

We need to start taking action domestically, because some degree of climate change is inevitable whether we like it or not. What I would say finally is simply this: that the nuclear threat and climate change are perhaps two greatest risks that humankind faces in our lifetime. In the case of nuclear threat, in the end it required citizens' actions campaign for nuclear disarmament etc. before something happened. I increasingly feel that in the case of climate change also if we were to leave it only to a dialogue between scientists and Governments it will not work. If we want action on this what we need is a global civil society movement which pressurizes Governments into taking actions which look beyond the short-term cost into the long-term implications.

Responses to Questions

What is your assessment of the outcome of the meeting of the Conference of Parties in Bali?

I am personally a little more positive about Bali than most people are. What happened in Bali? The Kyoto Protocol has hard obligations, which is to have a cap on emissions. In Bali the language got modified under U.S.A. pressure to "nationally appropriate strategies for involving commitments or actions". The U.S.A's. interpretation of this was that what we have agreed is that each country will announce what it is going to do. That's it, it is not something which I have to agree with you. All that is required is that each country should have a national strategy, and that it does not have to involve caps and commitments. It may be national strategy to say I am going to triple investments in renewable energy. It may be a national strategy which says that I am going to put some efficiency standards but it may not involve a cap. So the U.S.A. interpretation of this would be clearly resiling from the framework of Kyoto, and loosening the framework.

There is another phrase saying, "comparability of effort between countries and measurable, reportable and verifiable". The implications of this is that you do need to talk with others because otherwise how will you get the comparability with others and report to others about performance. Otherwise what does measurable, reportable, verifiable mean? So these disputes will go on, but it is a weakening of the Kyoto agreement as far as developed countries are concerned.

It is probably a strengthening of the agreement when it comes to developing countries, who earlier were not required to frame mitigation strategies. Under the Bali outcome they are required to frame mitigation strategies. But they have to be provided finance and technology.

So, I would see on balance the outcome is positive. At least you brought United State into the tent, and no global climate regime is credible unless Washington is part of it. They are the biggest polluter and there is no point in our having a climate regime if they are going to stay out. I also believe, no long-term climate regime is credible unless India and China entertain it. We have to be realistic on this.

Why is it that some people in Government are satisfied with weak global agreements?

Part of the explanation lies in the behavior of the Western countries particularly USA and George Bush. It is difficult to ask India and China to be more forthcoming if that gentleman in Washington says that there is no such problem. Now they don't deny it and today their thesis is no, but science will solve the problem. Between carbon sequestration and nuclear technology we have the answer and person no less than Jeffrey Sachs argued this or so he said a few days ago in the Tata Energy Research Institute (TERI) summit. He made rude remarks about George Bush and I felt like telling him.: "Why are you making rude remarks about George Bush. You sound exactly like George Bush yourself". Because that's exactly what George Bush says. That we do not need to adjust at the demand side. So part of the explanation for the stand in India is that there is this scientific hubris which says: that we will find a scientific fix for this, including, extraordinary things like deploying umbrellas in space to deflect the heat from the sun. There are other people who argue for simple fixes like convert the whole of Sahara into a Solar-Thermal power plant and all of humanity's needs for energy can be fulfilled. Scientists have habit of neglecting the social and economic and historical realities and seem to do most of their history at the back of an envelope.

Part of explanation is the normal dynamic of diplomacy which tends to look at success more in terms of did I get my paragraph in rather than in terms of the end result. O! I succeeded in getting that phrase shifted from the beginning of the paragraph. There was a phrase in the Bali text asking developing countries to take nationally an appropriate mitigation strategies in a measurable, verifiable and reportable manner subject to access to technology and finance. Our great contribution was, that we got the phase "measurable, reportable and verifiable" shifted from the beginning of the paragraph to the end, so that by implication it applied not just to the strategies but also to the provision of finance and technology. These are the little things which diplomacy thrives on. I am sorry. I am only teasing the Government!

How will climate action impinge on the poor?

The poor would be affected far more adversely by a failure to act on climate change. A failure to act will worsen water stress, health conditions, agricultural prospects and the condition of settlements. Who are the people who most likely to be affected by seawater intrusion? Who are the people who are going to be affected by water scarcity? By agricultural fluctuation? Malaria increases? The poor and among them women more than the men. So instead of worrying about what is the cost of trying to contain climate change to 2 degree centigrade, we should also ask what is the cost and who is bearing the cost if we don't do contain temperature change? Why do we treat doing something as a burden and neglect the cost? The costs of inaction are far larger for the poor than the costs of action. The rich would always look after themselves in these types of situations. The best anti-poverty scheme that we can have for the poor is to improve the security of their lives, not by handing out subsidies here and there. I would therefore strongly argue that Government has to address climate change because the greatest victims of climate change are going to be the poor not the rich. So it is very much in the interest of poor if we spend money for mitigation.

What would that be? The overall estimates of cost at global level have been made including some estimates for India. The maximum that we are looking at is 1% of global GDP. If we don't do anything the cost could go up to 5 to 6% of global GDP, i.e., a benefit cost ratio of 5 or 6 to 1.

What is the impact on women and health?

I don't know enough about the potential impact on women. As for health, one of the thing that will happen is that the areas in which malaria is endemic throughout the year would now cover virtually the whole country. At the moment, there are many parts of India where Malaria is not endemic throughout the year. This is one example of what temperature changes are likely to do in terms of health and so on.

3

Asean's Approach to Climate Change

P.V. Rao

Introduction

As rapid industrialization, premised on the growth philosophy, was set into motion by the Asian countries, it also unleashed serious concern about the ecological effects of industrialization. Over the years, the Asian countries – East and Southeast Asian – had become party to the alarming phenomenon of global warming and greenhouse impacts. The miracle economies of Asian countries and their growth models unexceptionally drew them into the global chain of environmental degradation and climate change. Growth leads to economic rise, competition, overproduction, market expansion, consumerism, employment opportunities and resource exploitation. But, unless the growth-inspired economic strategies are also inspired by the principles of equity and inclusiveness, they hamper social and natural balance. Unbridled industrialization and competitiveness leads to profit-making by few at the cost of many. The unregulated engines of growth and market forces alienate and marginalize several sections of society. The same forces disturb ecological balance by discharging greater carbon emissions through increased automobiles, high oil exploitation and consumption, conversion of farm lands into exclusive economic zones, deforestation and forced displacement of people. In 2006, Nicholas Stern, former chief economist, World Bank and advisor to the British government, cautioning about the economic threat from climate change in his Stern Report stated:

> It is quite evident that people want incomes to grow, they like to consume more and governments will respond to their demand for growth. The political challenge is to combine low carbon activities with growth. The challenge is not about cutting back on growth but finding a way for low carbon growth.[1]

Six major ecological damages caused by the above mentioned economic policies are: climate change, ozone depletion, agricultural intensification, deforestation, overfishing and mangrove conversion. Climate change is the most seriously debated environmental issue today by the international community. While both the industrialized and less industrialised developing countries share the concern, the former is mostly responsible for the serious ecological phenomenon. When calculated on per capita basis, high-income nations are responsible for 5.7 times more greenhouse gas emissions than low-income nations. The United States alone accounted for a 20.6% per capita share of global carbon dioxide emission in 2004, UK 9.8%, Italy 7.8%, France 6%, Germany 9.8% and the Russian Federation 10.6%. However the share of industrialized countries of Asia in carbon emissions is also registering an increasing trend. China's per capita emission share in 2004 is 3.8%, Japan 9.9%, Singapore 12.3% and South Korea 9.76.[2]

But poor nations bear the burden of climate change disproportionate to what they emit. When calculated on the basis of fossil fuel emissions, industrialized nations bear even a great share of liability, since they use more ozone-depleting substances. Though the low-and middle-income groups emit only 1.6 and 28 per cent of chlorofluorocarbon emissions, they suffer up to 15 and 44 per cent respectively of the ensuing health hazards. "Ecological damages are distributed more towards poor notions".[3] The following table explains the impact of climate change on the various ecosystems of the globe.

Ecosystem type	*Implication of climate change*
Wetlands	Variable water cycle, leading to inland wetlands to dry out with lower species diversity.
Coastal marshes	Habitat loss, impact on migratory species and flyway patterns
Forests	Major changes in vegetation, re-growth in new areas, significant losses through fires
Low-lying islands	Loss of land area / seabird nesting colonies
Arid/semi-and areas	Deserts likely to expand, become hotter and drier, loss of arable land
Coral reefs	Coral death even with minor rise in temperature, less growth due to increased CO_2
Mangroves	Decrease in area as coastal zone "squeezed" between sea and inland agriculture

(*Source*. *Down to Earth,* March 31, 2008).

Climate Change: concept and concerns

Traditionally, climate change was viewed as an environmental issue and did not figure as a key issue in the environment discourse of the economically dominant states of the North. Some of the advanced countries like the United States refused to subscribe to the climate change protocols passed at Koyto and Johannesburg. As the developing and underdeveloped countries realized the detrimental effects of the climate change patterns on their weaker economies and people including the coastal dwellers, the South articulated the subject as a major concern of development discourse. In otherwords, "climate change has metamorphosed from an elite, scientific, Northern issue into one which can usefully fit into the environmental justice agenda of the South."[4] Illustrious of the growing southern concern for climate change was the huge Rally for Climate Justice held at Delhi in October 2002 which coincided with the UN Climate Change meeting known as Conference of Parties 8 (COP8).

Food scarcity, drought, flooding, rain and water scarcity, denial of clean air, etc. are attributed to the excessive and alarming human disruption of the ecological balance. Sea warming caused by climate change "is reducing its (sea) ability to carry dissolved oxygen, potentially turning swathes of the world's oceans into marine graveyards." Researchers, Lother Stramma and Janet Sprintall warn that if global temperatures keep rising there could be 'dramatic consequences' for marine species and coastal communities.[5] In fact, a recent report by the Food and Agriculture Organization (FAO) warned the fishing communities of the adverse impact of climate change on their livelihood. In the developing countries, where about 42 million people are directly dependent on the fishing sector, climate change "will increase uncertainties in the supply of fish."[6] Thus, the "longer term concerns expressed by the IPCC (Intergovernmental Panel on Climate Change) and other expert documents about the impact of climate change do not evoke just the fear of more frequent and less predictable extreme weather, but the impact of increased greenhouse gases on the very ability of nature to sustain itself."[7]

Southeast Asian experience

The phenomenon of lop-sided and unequal growth, resulting in social inequities and global warming is evident in the Asian countries. Southeast Asia's "intensified economic activities, particularly in the industrial and agricultural sectors, and rapid urbanization in recent decades have resulted in the environmental deterioration of land and water resources of the countries concerned. In some instances, economic development has been associated with adverse impacts on the environment and resources, and have increased the income disparities between economic groups."[8]

In Myanmar, the Thai-built power project has led to the forced displacement of many people. Increased emphasis on agrofuel plantations, as palm oil in Indonesia, is reducing the traditional farming communities to poverty and causing environmental hazards. As an environmental activist representing *Watch Indonesia* observed: "Palm oil for biofuels is driving people into poverty. Largescale agrofuels plantations already are responsible for hundreds of human rights abuses. Furthermore, they have the potential to create more horizontal and vertical conflicts. This is not the climate justice that the people of Indonesia want to achieve."[9] Deforestation caused by the indiscriminate felling of trees by the timber industry and fuel wood consumers has disturbed the ecological balance in Southeast Asia. For instance, in Singapore large tracts of original forest cover was destroyed due to the use of firewood for extracting the juice from gambier. Similarly, timber extraction resulted in the virtual disappearance of precious teak wood from Myanmar and Thailand as also the depletion of forest reserves in the Philippines and Sarawak. Such largescale "destruction of forests interrupts the water and nutrient cycles, causing longer dry seasons and irregular river systems, enhances the carbon contribution to the atmosphere, enhances the loss of invaluable genetic material, and enhances the appearance of new pest and disease organisms."[10]

The ten member group of Asean countries, representing Southeast Asia and the Indo-China, are initiating individual and collective measures to deal with the environmental hazards caused by climate change. Both natural and material factors explain their resolve to deal collectively with the growing menace of climate change. While geography is mostly responsible for periodically damaging the human settlements and environment in the Southeast Asian region, the rapid industrialization and growth strategies adopted by the countries in the region are no less the sources of environmental degradation. Those who are familiar with the Southeast Asian topographical and maritime features can easily assess the high vulnerability of countries of the region to frequent cyclonic and environmental disasters. Most of these countries are situated in the tropics. Long coastlines, island and archipelagic composition, of Indonesia and the Philippines in particular, render them vulnerable to cyclonic storms, sea-rise hazards, tsunamis and the like. Their recurring sequence affect the ecological balance, coastal habitations, groundwater, fishing communities and the marine systems in general. Indonesia is the world's third largest emitter of carbon chiefly caused by largescale deforestation and degradation. It may be noted that next to Brazil and Congo, Indonesia has the largest forest coverage area in the world. The Heart of Borneo encompassing the highlands of Indonesia, Malaysia and the low lands of Borneo, habitat for many a wild animals in its vast forest area, is reduced almost to half its original coverage due to deforestation.

While the fury of natural disasters is the major source of ecological disturbances in Southeast Asia, global warming "also increases the frequency and intensity of tropical storms and induces more cardiovascular and respiratory diseases. Some parts of Southeast Asia are already experiencing a higher incidence of diseases such as dengue fever and malaria due to warmer temperatures."[11] Indonesia, world's largest archipelagic state straddled by numerous islands and islets, and the Philippines comprising over 7000 islands, are the frequent targets of sea-borne storms and floods. But these Southeast Asian countries are also, more recently, suffering from global warming and climate change. Not surprisingly therefore these two countries have been actively engaged in climate change negotiations since eighties.[12] In fact, the Philippines was active in launching the Intergovernmental Negotiating Committee at the UNFCCC (United Nations Framework Convention on Climate Change).

Singapore, not as concerned as her neighbours about the environmental effects, in recent times joined the regional initiatives to meet the climate change threats. As the city-state's contribution to the emissions is small, it did not feel compelled to take active interest in the Southeast Asian efforts to mitigate the effects of global warming on their societies. Thus Singapore, which earlier refused to sign the Kyoto Protocol, acceded to it in 2006. Perhaps the fact that Singapore was in the Chair of Asean's 13th annual summit meeting influenced her decision to conform to the Kyoto Protocol. It was Singapore which proposed 'Energy, Environment, Climate Change and Sustainable Development,' as the Asean's theme for the 13th summit held in November 2007, just about one month ahead of the UN Conference on Climate Change held at Bali, Indonesia, in December 2007.

The Singapore Declaration

In addition to their participation in the global initiatives on climate change, all the ten member countries of the Asean group are equally active in mooting regional strategies to combat the challenges posed by the rising phenomenon of global warming. At the 13th Asean summit held at Singapore the ten members of the group along with its summit partners – China, India, Japan, South Korea – signed on November 20, 2007 the Declaration on Climate Change, Energy and the Environment, commonly called the Singapore Declaration. The Singapore Declaration was preceded by the Asean Declaration on Environmental Sustainability and the group's adhesion to the APEC Leaders' Declaration on Climate Change, Energy Security and Clean Development adopted in September 2007.

Asean's Singapore Declaration recognizes the new challenges posed to the region by the rapid economic development, growing urbanisation and the rising energy consumption. It called for the need for environmental management in view of the projected doubling of Asia's 1.7 billion urban population between 2000 and 2030. Hence it is necessary to devise an effective approach to the "interrelated problems of climate change, energy security and other environmental and health issues…." Committing themselves to the goal of stabilizing atmospheric greenhouse gas concentrations, the Asean countries laid out the following mitigation and adaptation strategies to minimize the impact of climate change.

- Encourage the deployment of clean technology in the region through various means, such as investment, technical and financial assistance, and technology transfer.

- Commissioning of national and where appropriate, joint studies to assess the impact of climate change and environmental protection efforts within the region.

- Enhancing regional cooperation to develop cost effective carbon mitigation technologies, cleaner fossil fuel technologies including clean use of coal, and to produce environmentally-friendly and sustainable bio-fuels.

- Cooperating for the development and use of civilian nuclear power.
- Promote cooperation on afforestation and reforestation, and to reduce deforestation, forest degradation and forest fires.
- Continued support for UNFCCC work to stimulate action to reduce emissions from deforestation in developing countries, including through appropriate international incentives and assistance.[13]

The lengthy Declaration laid the roadmap for an environmental friendly regime in Asia by identifying almost every challenge posed to the eco-balance in the region as well as devising the needed strategy to combat such threats. The same spirit and commitment by the Asean group was clearly evident at the UNFCCC meeting at Bali in December 2007.

Asean and the Bali Conference

On the eve of the Bali Conference Asean ministers in charge of environment met to debate the regional position on the issues of climate concern to be discussed by UNFCCC. In an unanimous statement released to the press, the ministers recognized that "climate change poses serious risk and challenges to developing countries, and called for urgent global actions and responses to address climate change, on the basis of the principles of equity, flexibility, effectiveness and common but differentiated responsibilities....." They also agreed, in view of the region's high vulnerability to the adverse impacts of climate change, to encourage efforts to develop an ASEAN Climate Change Initiative (ACC) envisaged to be a "consultative platform to further strengthen regional coordination and cooperation in addressing climate change, and to undertake concrete actions to respond to its adverse impacts. The Initiative will strengthen the region's capacity both in mitigation and adaptation efforts, and in particular to bring forward the region's interests and priorities onto international negotiations on future climate regime as appropriate."[14]

The Bali summit launched a two-year negotiation phase to devise a new framework replacing the Kyoto protocol in 2013.

The United States, which refused to so far to sign the Kyoto protocol, agreed to join the next two years of negotiations, heralding thus a new phase in the global effort to mitigate the greenhouse impact. The Bali agenda for the next two years, 2008 through 2009, would largely focus on four issues of related concern: mitigation, adaptation, financing, and technology.

Conclusions

Beginning with the decade sixties of the last century, Southeast Asian countries had attained fairly higher standards of economic growth in a decade's time. Gradually, these countries also began to realize the impact of capitalist mode of development on rural populations, natural resources and the environment in general. As the region continued to attain rapid growth rates and industrialization, resulting in the 'miracle economies,' ecological impact of such rapid economic transformation drew the attention of governments and social groups. While national initiatives to meet the challenges of the growth economies could not be ignored, the development of Asean as the regional collective group elevated the eco debate to regional level. The original debate on the growth versus environment steadily shifted to concerns over climate change. At the same time, global concerns and debate over climate change too influenced the regional debate. It remains to be seen how and to what degree the Asean countries will reconcile the regional concerns over the effects of climate change with their subscription to the global conventions.

References

1. *Down to Earth,* November 30, 2007.

2. See, "Carbon Dioxide Emissions and Stocks", *Economic and Political Weekly,* February 2, 2008.

3. *Down to Earth*, April 15, 2008.

4. Timothy Doyle, *"An agenda for environmental security in the Indian ocean Region," Sanjay Chaturvedi and Dennis Rumley*, Ed. Geopolitical Orientations, Regionalism and Security in the Indian Ocean, South Asian Publishers, New Delhi, 2004, p. 165.

5. See, "Oceans' death zones spread," *Times of India*, May 20, 2008.

6. See, "*Cut emissions to save fish,*" The Hindu, July 29, 2008.

7. Peter Stoett, "A Climatic Normative Catalyst: Climate Change and Global Environmental Justice," *The Icfai University Journal of International Relations, Hyderabad,* Vol. ii, No. 3, July 2008, p. 59).

8. See,Colin MacAndrews and Chia Lin Sien, Developing Economies and the Environment, the Southeast Asian Experience, McGraw Hill, 1979, p. 11.

9. See, Bali talks, NGOs from North and South challenge catastrophic climate and ecosystem impacts of large-scale bio-fuel Production *http://ec.europa.eu/environment/climat/home_en.htm.*

10. Colin Mac Andrews and Chia Lin Sien. p. 100.

11. Hamilton-Hart, Natasha, "*Singapore's Climate Change Policy: The Limits of Learning,*" Contemporary Southeast Asia, Vol. 28 *http://muse.jhu.edu.journals.*

12. Colin MacAndrews and Chia Lin Sien, p. 110.

13. See, ASEAN Secretariat, *Singapore Declaration on Climate Change, Energy, and the Environment,* November 2007.

14. *"ASEAN Cooperates on Climate change,"* Press Release, *http://ec.europa.eu/environment/climat/home_en.htm*

4

Impact on Indian Ocean Region

S.K. Chawla

"The issue of Climate Change is one that we ignore at our own peril. There may still be dispute about exactly how much we are contributing to the warming of earth's atmosphere and how much is naturally occurring, but what we are scientifically certain is that our continuous use of fossil fuels is pushing us to a point of no return and unless we free ourselves from dependence on these fossil fuels, we are condemning future generation to global catastrophy"................

(Barack Obama)

The only comment on the above quote to which most of us may agree is *"Look who is talking"?* A change which must occur in double quick time.

Both IPCC and HDR have agreed, rather reiterated, that global warming is an established fact, and that it is due to human intervention. There are recommendations on how to mitigate and adapt to the likely impacts of this major worldwide phenomenon. On the region-specific canvas of South Asia, the issue of energy security (in general terms) emerges as a high priority area (India's offshore oil and gas fields in western and eastern coast regions may be cited as a specific illustration).

The socio-political and economic disparities between developing and developed nations offer an opportunity for challenging the threat of inaction coming from opinion holders with digressing tendencies. Does it not seem that the delay in implementing the recommendations of Kyoto, Rio and Montreal is likely to create impediments on renewed thinking? Is it not desirable to consider developing techno-cultural and socially acceptable mandates to maintain/accelerate economic growth in the region without increasing the conflicting and self-defeating developmental activities of diverse economies?

Energy, being the main economy driver, is a dire necessity for the very existence of humanity. Its current-day, largely fossil-fuel dominated production and usage, however, represents the

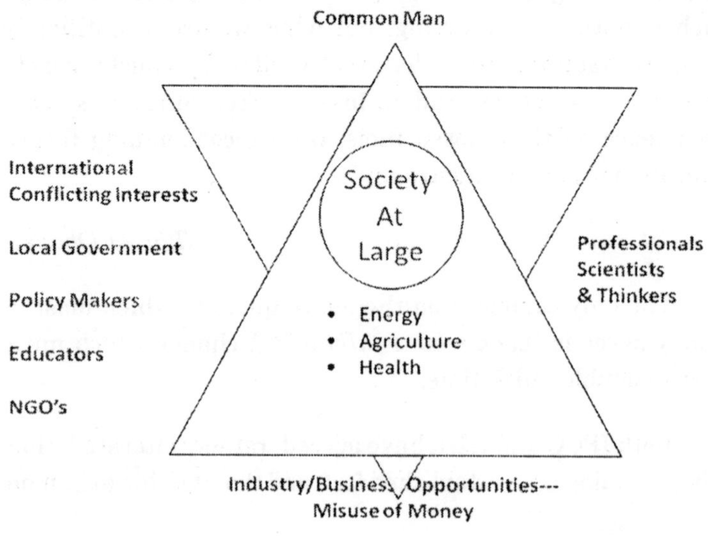

primary cause for the increasing temperatures across the globe. Thus, the impacts of climate change — the melting of glaciers, the rise in water levels of oceans, the increased frequency of storms, tsunamis, etc. — result, for the main part, from the greenhouse-gas (CO_2) emissions associated with the burning of fossil fuels.

Populations dwelling in low lying and coastal areas are thus going to be increasingly subjected to immense sufferings resulting in migration, demographic changes, social pressures, cultural imbalances and distortions, and new political and economic orders. Let us look at the play-ground and the players in the game. See the above simple-looking diagram is in fact functionally quite intricate and tends to lose focus on the central object, viz. *the common man at the grass-roots level* (gainer or loser). Where do we see MAN: at the bottom, at the top, or nowhere at all within the central orbit?

The next question arises in relation to the "hexagon" shown below. This depicts the total system as an elementally differentiated and loosely integrated dynamic body with dissipating driving force or steam Can this be channelized so as to become a self-driven socially committed entity? The answer is not simple.

"....The heat trapping gases we send into the atmosphere in 2008 will stay there until 2108 and beyond. We are therefore making choices today that will affect our own lives, but even more so the lives of our children and grandchildren. This makes climate change different and more difficult than other policy challenges...."

With this back-drop, let us try to address the issue as to whether future energy development for economic growth is bound to continue impacting climate change or whether it is climate change *per se* which will determine the extent and

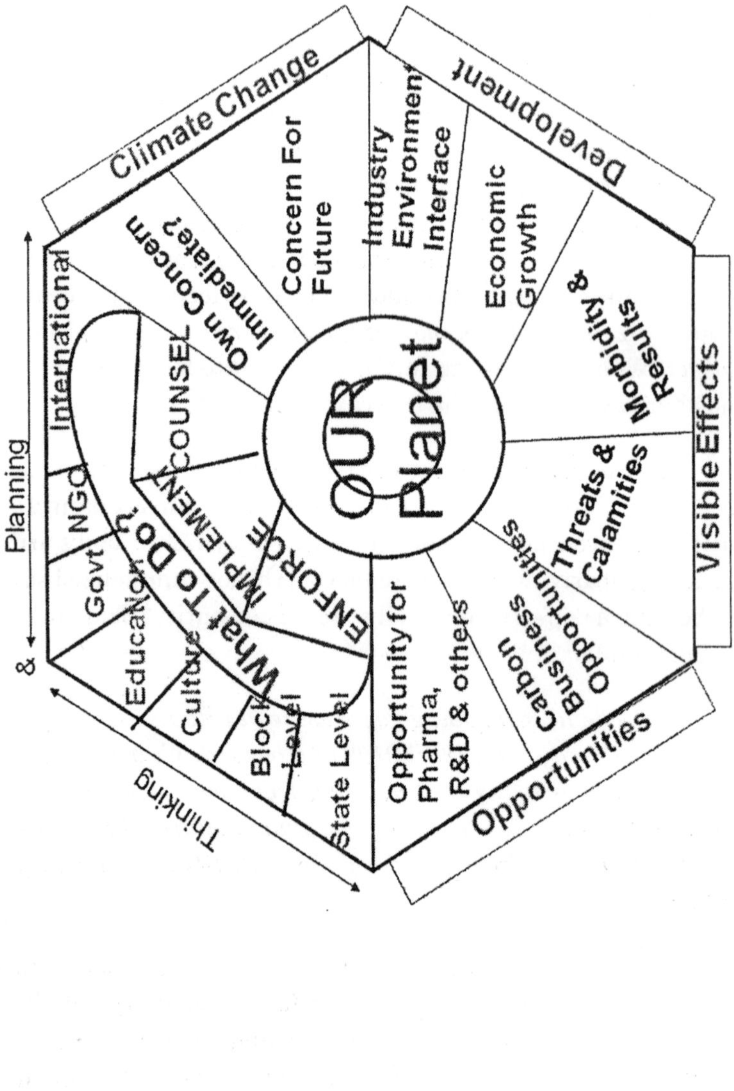
Climate Change
Development
Visible Effects
Opportunities
Thinking
&
Planning
OUR Planet
Concern For Future
Industry Environment Interface
Economic Growth
Morbidity & Results
Threats & Calamities
Carbon Business Opportunities
Opportunity for Pharma, R&D & others
State Level
Block Level
Culture
Education
Govt
NGO
International
Own Concern Immediate?
COUNSEL
IMPLEMENT
ENFORCE
What To Do?

modality of energy development. I see things as viciously interlinked. There is little doubt that global energy requirements will continue to rise sharply. It is only widespread international cooperation that can allow this to happen while at the same time curbing the current rate at which the climate is changing.

The unexplored areas and way forward

There is no simple recipe, but it is clear that energy savings measures (e.g. through increased efficiency in energy utilization practices) must be complemented by the large-scale implementation of novel CO_2-free technologies for energy production. The developed countries should set the example by increasingly replacing the use of fossil fuels by the large-scale deployment of the "new renewables" such as solar and wind. The relatively high costs are the major deterrent for this at the present time, but the time has come for bold measures to be taken. These should include the increased worldwide use of nuclear energy — a commercially well-established, CO_2-free technology that, through expanded deployment, could help alleviates challenges in both electricity and transport sectors.

Let us see whether the oceans, with which our societies are intimately linked, offer any opportunities or scope to address the bigger issues mentioned above. I hope we all agree that:

- Oceans belong to all nations and, other than their own economic development zones, the oceans' wealth is for everybody.

- Violations and perturbations in one part of the oceans affect everybody, e.g. pollution through oil-spills destroys fisheries, undiscovered marine flora and fauna, etc. Ocean currents such as the Gulf Stream have no political allegiance and are therefore everyone's friend or foe.

- Potential energy resources related to the oceans (other than petroleum), e.g. waves, currents, tides, wind, overlying solar radiation, geothermal, etc. are not the property of a single nation or individual. Everyone has the right to channelize the corresponding energy for economic use. As indicated above, these are areas of future frontier technology and should provide the framework for concerted international research, as well as cooperation in economic development.

- The amount of research being done in Energy related fields especially in "Re-Cyclable" resources should be sponsored by International Development Institutions as the advantage in the long run is global.

While the scientific community examines the new frontiers, the present generation must see the rationale in protecting society against the ill effects of accelerated development (e.g. resource abuse, waste mismanagement, inefficient energy usage, etc.) and learn to share the burden of poverty/scarcity. One must discourage the vulgar display of riches via measures such as rationalization of the transport sector and development of infrastructure on an equitable basis. Furthermore, knowledge and discipline need to be imparted to the young, so that energy savings and high energy-usage efficiency become part and parcel of future social behaviour. Indeed, a drastic transformation in our life styles and habits is called for if technological developments — such as the widespread use of renewable energy, water harvesting, sewage management and recycling — are to have their desired effects.

Can we learn some lessons from the USA, which buys oil from the Middle East and conserves its own, or from Singapore and Japan, which barter technology for raw material and use value-addition methods to accelerate their economy? Why should we burn coal which needs to be transported (with 40% ash) to distantly located power stations, or spend precious time on futile discussions in air-conditioned offices, rather than taking corporate initiatives on setting up novel types of pilot plants and stand-alone energy efficient systems? Can we not — as responsible individuals, communities and corporate organizations — be more generous in inputting a small percentage of our time and money into a *corpus*, set up exclusively for the global well-being of our society, rather than pointing fingers towards governance systems on which we have little control?

Leads Through Social and Cultural Arsenal

It can be stated with reasonable confidence that the concern for future well being is visible on the social face of humanity today, though it is the culture of consumerism which needs to be withdrawn to a certain degree in our way of life.

Parallel actions are needed on the social front *per se*. Thus, in addition to meeting the scientific and technological challenges of global warming and climate change, we also need to put our heads together on how best to achieve the warming of hearts and the change of social values. Let us constitute small groups of intellectually likeminded people, promote awareness programs on energy, environment, health and culture, then connect them to corporate organizations, government, media, spiritual organizations, academia, etc. Public opinion leaders may take the role of resolving dissidence. Let the action start at the level of grassroot viz locality level – welfare societies, religious and spiritual organisation, Municipalities, city corporations and the likes. At the school and college level the issue of healthcare through environmental protection and upgradation should become a part of curriculum.

Corporate organisations need to be actively involved in the process of imparting knowledge. They can do this through incentive schemes, based on their budget for development and social services, even if they do not contribute directly to the financing of new projects.

Lastly, let us call upon the various governmental ministries to be involved in collaborative ventures with the private sector to launch movements on clean development and green energy.

This will go a long way for health care, although may like regression of modernity respect for future generation and young who need water, organic food, pure air, and disease free body can only come if we optimize Thermal – Hydro balance of power, Encourage usage of Alternate and Renewable Energy in isolated areas, Reduce pressure on Grids, Go for Hybrid modes of energy usage, and in fact bring a transformation in our living styles and habits. Development of Energy efficient buildings, Water Harvesting, Sewer Management and recycling are some other areas which need attention.

In this connection the concluding remarks of Indian Planning Commission's report titled "Integrated Energy Policy 2005" are noteworthy.

".... India faces an enormous challenge in meeting its energy requirement over the coming 25 years to support a growth rate of 8 percent. This challenge can be met with a coherent approach which develops all her energy resources. The main areas, action for which detailed policy recommendation are made as follows:

- Promoting Coal imports.
- Accelerating power sector reforms.
- Cutting cost of power.
- Rationalisation of fuel prices to mimic free market prices that promote efficient fuel choice and substitution.
- Promoting energy efficiency and conservation.

- Augmenting energy resources and supply.
- Encouraging renewable and local solutions.
- Enhancing energy security.
- Promoting and focusing Energy R & D.
- Promoting energy security through entitlements for the poor, gender equity and empowerment.
- Creating an enabling environment and regulatory oversight for competitive efficiency.

With the recommendations of the Committee, India can meet her energy requirements in an efficient, cost effective way and be on a path of sustainable security...."

Let the process begin and come on a formal organisational track with clearly defined functional boundaries.

References

1. HDR Report 2007 – 2008.
2. IPCC Report.
3. *www.iea.org*
4. Integrated Energy Policy (Indian Planning Commission).
5. Several other articles from the internet.

5

Technologies For Carbon Dioxide (CO_2) Sequestration

Malti Goel

1. Introduction

Climate Change Will Devastate South Asia - Daphne Wysham and Smitu Kothari in Hindu 18th April 2007

Climate change 'ruining' Everest – Richard Black, BBC environment correspondent.

Lohachara Island in India's part of Sunderbans is washed off in the rising sea caused by global warming— The Independent 24 Dec. 2006.

Global warming threat and consequent climate change are becoming household words in the 21st century. The projections of future impacts of climate change in different parts of the world are devastating as evident from the above quotes.

South Asia and South East Asia are among the regions to face much damage in future. On the other hand, these regions are energy starved, having world's lowest per capita consumption, yet much of their energy comes from fossil fuels and fuel wood. Coal contributes to almost 60% in India and 40% energy in South East Asia.

Rapid industrialization and high energy consumption in industrialized countries during past century has disturbed the natural carbon cycle. For meeting their energy demands carbon dioxide has been increasing in the atmosphere from the fossil fuel burning. It is expected that emission of carbon dioxide (CO_2) in the atmosphere will continue to increase, as major population of the earth has no access to electricity yet. In this context, sequestration of excess CO_2 has emerged as a technology solution for allaying climate change impacts. It simply means capturing excess CO_2 from the atmosphere and fixing it in such a manner that it is permanently stored away from the atmosphere. The technologies are R&D intensive and costly at present. This paper describes emerging technologies of CO_2 sequestration with particular reference to India and South East Asian countries. As these countries are surrounded by long coastal lines, implications and critical issues associated with marine sequestration of CO_2 are also discussed.

2. Climate of the Region

The climate of India and South East Asia has some similarities and some dissimilarity. Most of the region in South East Asia and part of India have almost uniform temperatures and high humidity throughout the year and the climate is characterized as monsoonal.

2.1 India's climate

India has 7000 km long coastal line on its three sides. The Himalayas in north have an important effect on the weather and climate. The climate zones are divided in Monsoonal, Dry and hot temperate zones. There are four main seasons as winter, pre-monsoon, monsoon and post-monsoon. The ocean plays an important role. The lowering of sea surface temperature over Arabian sea from 28 °C to 25°C resulting from Somali current triggers the summer monsoon.[1] Extensive meteorological research studies [2,3] have been carried out on the Indian ocean region to understand onset of monsoon, inter-relationship of ocean and climate, storm surges, cyclones etc.

2.2 South East Asia

Southeast Asia comprises of about a 5000 km long chain of 20,000 islands strung between Asia and Australia (Fig.1). The most of the region has almost uniform temperatures throughout the year, high humidity and their climate characterized as monsoonal rain. In the South East Asian region changing seasons are more or less associated with rainfall variation rather than with temperature variations. There is, however, a high degree of climatic complexity within the region. In most parts tropical moist climate has led to immensely high bio-diversity and large regions of rain forests.

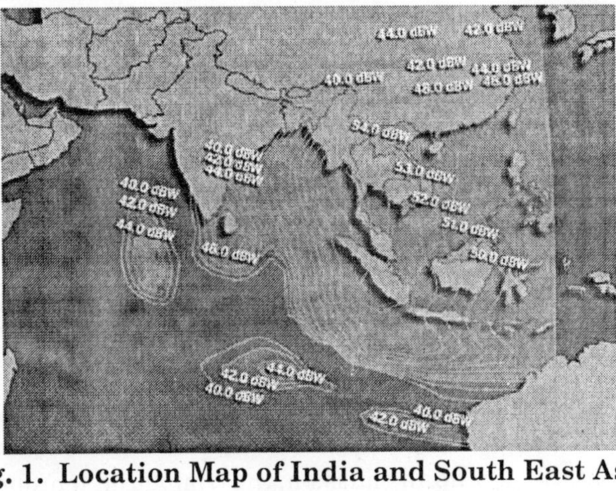

Fig. 1. Location Map of India and South East Asian Region.

3. Climate Change Context

Global warming is a complex phenomenon, not fully understood. The fourth Assessment Report of Intergovernmental Panel on Climate Change (IPCC) in 2007 has concluded that global warming is real and measures to mitigate it are urgently needed. Its adverse impacts have been seen to result in triggering melting of the Polar Ice Caps, recession of glaciers in temperate regions and a slow but inexorable rise in sea levels. It has been said that on its own global warming can lead to many more disastrous impacts on the Planet earth, its climate and extreme events. The report lays shocking scenarios for India and the rest of South Asia. Bangladesh is slated to lose the largest amount of land globally equivalent to approximately 1,000 square km of cultivated land and displacing millions of people living in coastal belts. It is projected that climate change as a result of global warming will cause an overall deterioration of growing conditions for crops and natural ecosystems in developing countries.

In this context, the only global treaty that exists has been the United Nations Framework Convention on Climate Change (UNFCCC) signed by 192 countries to stabilize greenhouse gases in the atmosphere and address climate change issues. Among the major and minor greenhouse gases identified, carbon dioxide (CO_2) is most important having a contribution of more than 50 % to the total warming in the Earth's Atmosphere. The regional distribution of CO_2 emission for fossil fuel emissions are depicted in Fig. 2. Under the UNFCCC the concern for global warming has attracted attention of global scientific community to address the challenge.

Fig. 2. CO_2 emissions from fossil fuels combustion and land-use changes. (width of the bar represents the population and height the per capita emission, the triangle on top suggests land use change. Source - IPCC 1995.)

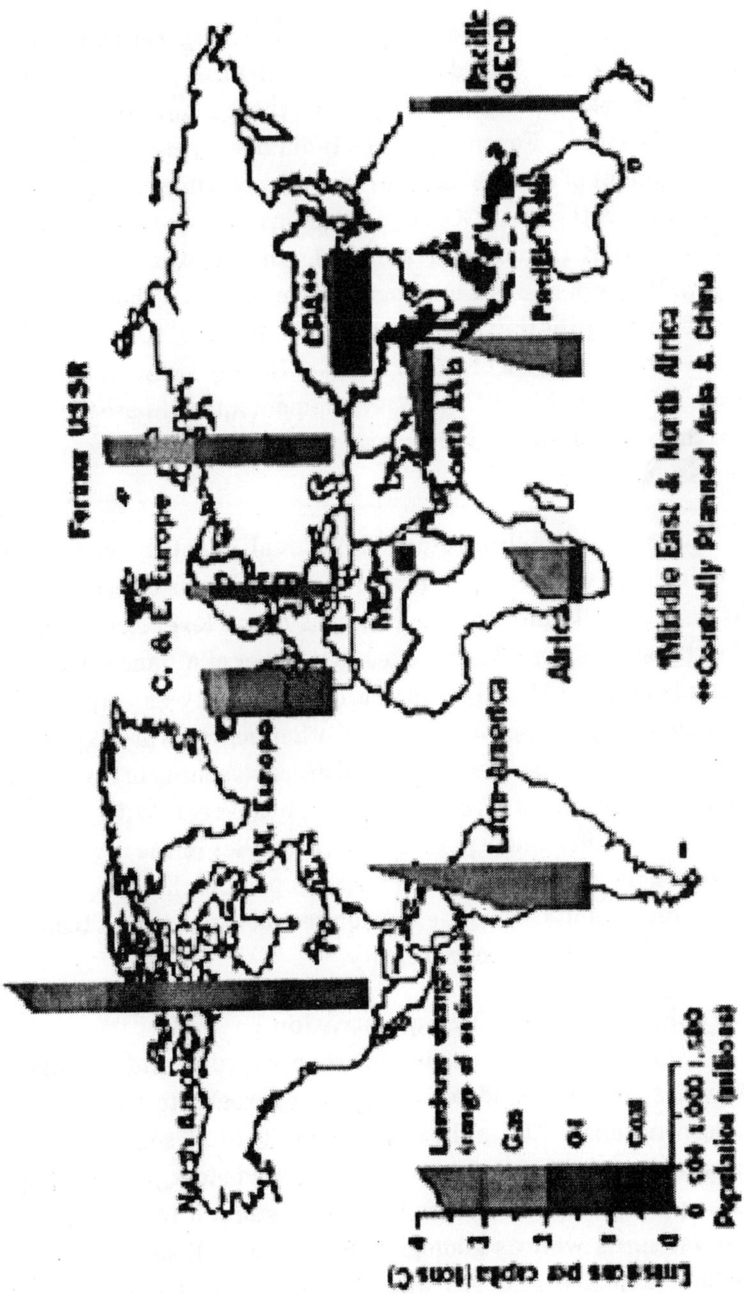

Former USSR
C. & E. Europe
W. Europe
North America
Latin America
Africa
Pacific OECD
Pacific Asia
CPA**
*Middle East & North Africa
**Centrally Planned Asia & China
Gas
Oil
Coal
Emissions per capita (tons C)
Population (millions)

In these times of rapidly developing technologies, voluntary action to adopt low carbon intensive, economic development could provide the path for these countries to remain in the forefront. For many years India has already focused on development of renewal sources of energy and improving energy efficiency. A thrust to CO_2 sequestration research by identifying the areas of R&D has been launched.[4] India is also signatory to Kyoto Protocol and is a member of technology collaborations in Carbon Sequestration Leadership Forum (CSLF) - an International Forum of Department of Energy USA, and of Asia Pacific Partnership in Clean Development and Climate (AP6), a partnership of six nations contributing to 75% of excess CO_2 emissions.

As Small Islands and low-level coastal areas are particularly vulnerable to sea level rise millions of people are at risk from climate change concerns of sea level rise. In the South Asian regions 36 countries as member of Alliance of Small Islands States (AOSIS) face sea level calamity as a result of climate change. Their tourism and other business are expected to the affected by shrinking of land area, bleaching of coral and killing of tiny creatures that replenish the reef. With 40% of their population below poverty line, countries in the region are expected to suffer greatly the adverse impacts of climate change. Preparedness in CO_2 sequestration technologies would lead to sustainability and energy security.

4. Technologies of CO_2 Sequestration

The CO_2 sequestration (carbon capture and storage) provides scientific and technological approach to reduce net concentrations of CO_2 in the atmosphere. The CO_2 sequestration encompasses all form of carbon capture and storage, including its capture from the atmosphere and storage in terrestrial ecosystem as well as geological formations. Essentially the carbon capture and storage technology has three main components, starting with capture of CO_2 in the atmosphere emitted from large point sources; fixing it or transporting it to a possible location where it can be safely stored; and finally processed and fixed.[5]

4.1 CO_2 Capture

The CO_2 can be captured from the flue gas of industrial / power plants or from the natural gas. At present 8100 micro and macro CO_2 point sources including power plants and industrial units have been identified around the world. There are comprising mainly fossil fired power plants and total emission are approximately 15 Gt CO_2 annually. CO_2 separation technology has been developed and its application has been tested. From a coal combustion based power plant/ CO_2 capture technology as end–of-pipe solution, is based on chemical absorption, membrane separation, physical adsorption and cryogenic separation methods. From natural gas the CO_2 capture technology is grouped in Post-combustion and Pre-combustion technologies. As of now there are many uncertainties and severe penalties in adoption of post combustion as well as pre-combustion carbon capture in a large-scale plant. Captured CO_2 from the atmosphere is absorbed and compressed. The CO_2 liquefies under critical condition of temperature 23°C and high pressure. It can be transported at 100 bars from the capture point to possible geological and terrestrial storage sites. The CO_2 capture means significant addition to the cost of electricity and great deal of R&D efforts are being made to make it cost-effective. R&D initiatives in India for further research in carbon capture material and process development has been targeted.

4.2 CO_2 Storage

Storage of CO_2 can occur in territorial land and water within the stipulated regulations of the regions. Oil and gas on -shore and off- shore empty fields, un-mineable coal seams, aquifers and terrestrial forests form potential storage sites for sequestration of carbon dioxide captured from large point sources. Underground storage potential as estimated from studies conducted by International Energy Agency Greenhouse Gas R&D is depicted in Table 1. Among these, natural gas fields have seen to have largest storage capacity. The oil fields can provide an opportunity for Enhanced Oil Recovery (EOR) using CO_2 injection to displace the left over viscous oil. In India it is proposed to inject CO_2 captured from a gas plant for EOR into Ankleshwar oil fields after secondary recovery methods have been tried.[6] Other active storage option for CO_2 capture is in the coal beds. It is expected the enhanced coal bed methane (ECBM) process can increase the conventional methane recovery of around 40 % to 90%. However, these options are in research stage and no demonstration units exist as yet.

The processes of EOR and ECBM present an opportunity for business from CO_2 storage.[7] On the other hand permanent fixation of CO_2 in rocks by reacting pulverized alkaline rocks with CO_2 captured can give insights to understanding the natural geological processes of rock formation. In India scientists at Geophysical Research Institute in collaboration with Pacific Northwest laboratory, USA envisage feasibility studies on CO_2 trapping mechanisms of geological storage in Deccan Continental flood basalt rocks province at a depth greater than 800 m. Storage in deep saline aquifers is another CO_2 storage option. These

Table 1: Storage capacity of CO_2 in possible underground location and the cost per ton*

S. No	*Probable storage site*	*Potential CO_2 storage capacity in Gt*	*IEA Estimates of storage in \$/t*
1.	Depleted oil fields	126	Net cost saving
2.	Depleted gas fields	800	7-17
3.	Unmineable coal beds	150	20
4.	Deep saline formations	400-10000	5-17

repositories would not yield any value-added by-product, however need to be pursued as research problems as the gap in knowledge exists to know about CO_2 mixing with water, reaction with adjoining rocks, increase in pressure and induced seismic risks in the vicinity. India's Carbon Capture and Sequestration policy does not consider underground storage of CO_2 for safety reasons. Technology needs to be developed and tested in industrialized countries before implementation in developing countries.

Terrestrial biosphere is estimated to sequester large amount of carbon dioxide and is thought to be one of the most cost effective means of reducing atmosphere levels of CO_2. Plants are natural reservoirs of CO_2 and CO_2 fertilization effect leads to better growth of plants. Terrestrial storage would also result in other benefits like value addition in forestation and enhance crop productivity. It can lead to development of advanced crop species and add to cultivation strategies. There is a significant potential in South Asian Countries for terrestrial sequestration of CO_2 which needs to be studied. Enhanced CO_2 absorption rate in vegetation and cropland are expected through CO_2 fixation. Wetlands, mined or un-mined forest sites and high way construction sites can also act as CO_2 sinks.

New possibilities exist in bio-fixation of CO_2 from industrial waste gases through microbial and micro-algae based processes. Technologies are envisaged for production of value added products and fuel, through genetically modified micro-organisms. A number of viable alternatives for CO_2 sequestration apart from natural terrestrial ecosystem are being explored worldwide. Development of elemental carbon, polymeric carbon and carbon nano-fibres as well as nano-material compositions, are industrial ways to utilize captured carbon dioxide in futuristic materials research. Some of the research initiatives in India[8] for terrestrial and ocean storage include :

- Enhanced photosynthesis to capture CO_2 in different plant species
- Micro-algae and Microbial Fixation of CO_2 from Industrial effluents for value added products
- Terrestrial and agro-forestry modeling studies to enhance sequestration rate of CO_2 in ecosystem
- Regenerative carbon dioxide removal by chemical and biological means
- Absorbent and adsorbent material development for CO_2 capture
- Adaptive response of micro algae to high temperature CO_2 coupled with reduced pH
- Studies on enhanced recovery of hydrocarbons from gas hydrates by CO_2 injection and capture

5. Marine CO_2 Sequestration

Oceans environment is largest global reservoir for CO_2 containing almost 4,000 gigatons of CO_2 in organic and inorganic forms, as compared to 750 gigatons in the atmosphere, of which 610 gigatons is in the terrestrial biosphere. Marchetti[9] first conceived the idea of injecting CO_2 to deeper depth of Atlantic Ocean for mitigating excess CO_2 from the atmosphere. It is believed that sequestration of CO_2 can be planned for its disposal on the sea surface using direct dispersion from the ship. In liquid form, it could also be brought to appropriate sequestration site through ships or a pipeline and could be injected to different depths up to the sea floor. Introduction of liquid CO_2 at the surface or up to 300 meter depth has been proposed to make it cost- effective. However, marine water acidity increases thereby affecting the marine life adversely and chances of it returning to atmosphere through surface plumes become high. Assuming that the injection is made at a greater depth below 1500 meters, then liquid CO_2 can be expected to form a lake on the sea bed. At a depth of 3000 m, pressure is 300bars and temperature is about -4°C. The CO_2 would remain in liquefied form and being more dense than water would remain trapped for a longer time. Disposal of blocks of frozen CO_2 into sea bottom has also being considered in Thermohaline zones. It would however draw still higher energy and environment penalty, and are not recommended. Various options of CO_2 in marine sequestration are shown in Fig. 3.

CO_2 disposed in ocean would require a study of interactive physical, chemical, biological and geological processes. Modeling studies predict increased acidity from CO_2 injection and change in pH. There are about 5000 species of marine phytoplankton, which serve as a base of the aquatic feed web, providing an essential ecological function for all aquatic life. Deeper injection of CO_2 into ocean is likely to affect marine species over a wider area unless it is contained in a limited area. Proof of concept experiments are being planned for studying physics of disposal of CO_2 at depths of 1000 meters or more and studying physical and chemical impact in the marine ecosystem.

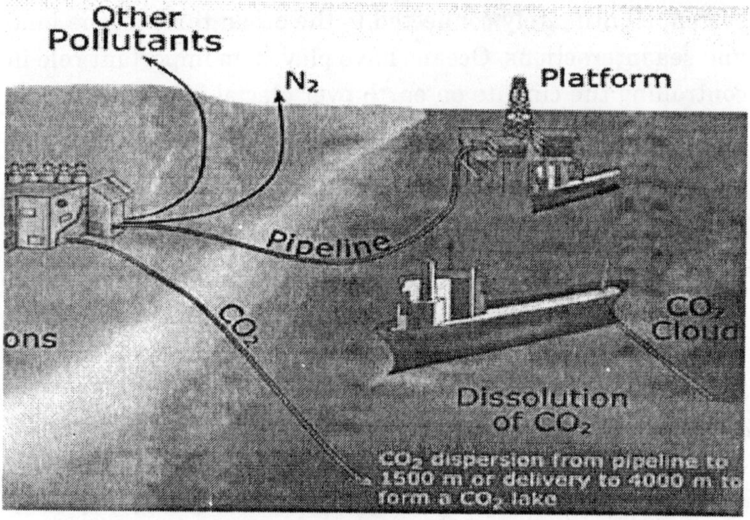

Fig. 3. Ocean sequestration of CO_2. Source- Princeton University.

Other option of marine sequestration is injection or disposal of CO_2 in the geological formations beneath the seabed. Such a possibility has been tested by Norway in the North Sea. Starting from 1996, about one million tons of CO_2 has been injected in the sandstone aquifers and continuous monitoring is being carried out. Permanent solution for pumping CO_2 into porous sediments, a few 100 m into sea floor in deeper parts of ocean > 3000 m for storage has been proposed. Beneath sea floor, CO_2 would interact with minerals in the surrounding fluid form to produce hydrate crystals, which would plug the rock pores and act as secondary cap. It may however, dissolve gradually over a long period of say 100 years depending on the physico-chemical reactions. In that case it may have potential of leaking by diffusion in a very slow process, that may take thousands to millions of years.

Little is known about the consequences of CO_2 sequestration on natural biogeochemical cycles.[10] Although ocean and atmosphere time-scales are very different, the climate system is intimately connected to the ocean through the land and sea interactions. Oceans have played an important role in controlling the climate on earth over glacial-interglacial time scales through CO_2 exchanges. The air-sea exchanges, turbulence and local mixing may lead more rapid injection in largely unpredictable manner. Biological pump continuously transfers CO_2 from the ocean surface to greater depths as warm water at the surface can hold much less.[11] When deep ocean water rises to the surface over a long period, water gets heated and releases CO_2 to atmosphere. Phyto-plankton have a tendency to sink into the deep ocean, by the force of gravity. They also pull carbon from upper ocean to greater depths where bacterial and other organisms metabolize and release it back into water as CO_2, enriching CO_2 in the deep ocean. Biological pumps also act to transport organic matter to deeper areas and strips nutrients such as silicate and phosphate from the surface layer.

CO_2 concentrations in deep waters of Arabian sea and Bay of Bengal is highest in the world. However, bio-geo-chemical cycle of CO_2 changes and how excess CO_2 would affect the climate/ weather circulations patterns is not fully understood.

5.1. CO_2 fixation through Iron Fertilization

Ocean planktons contain only 3 gigatones of CO_2 as compared to terrestrial plants, which have about 610 gigatones. However, regeneration of biologically produced carbon materials in the Oceans is faster than on the land. Presence of iron particles further promotes CO_2 absorption in phytoplankton's for growth. Hence sequestration of CO_2 through biological fixation on iron fertilization of ocean areas could enhance the ocean capability to act as CO_2 sink. Such experiments have been performed in the North Sea. Iron fertilization experiments conducted during 1993-2005 have suggested that addition of iron did not enhance productivity, but it may tend to increase it with time. Iron enrichment enhances nitrate and urea uptake during initial phase as compared to ammonium uptake, but in the later phases ammonium uptake is also seen to increase. It is believed that insufficient availability of nutrients such as iron on surface water can lower productivity in these regions.

By one estimate enhanced biological fixation using fertilizations on ocean areas with nutrients is estimated to consume 4,70,000 tons of iron per year to be spread over 25% of the Ocean surface, repeated for indefinite period. In addition, it could lead to deoxygenating of water column over a wide area and cause fundamental changes in the marine planktons species as well as result in enhanced bio-organic production of N_2O, which is another global warming gas. Whether long-term absorption of CO_2 will occur or increase in other greenhouse gases (methane and nitrous oxide) may actually result in increasing plankton boom. Studies have be concluded on the measurement of CO_2 emission in different parts of the Oceans.

South Asian region emits 0.72 Gtc/ year.[12] While the Arabian Sea emits 0.189 Gt/ year, the other arms of North Indian Ocean i.e. Bay of Bengal actually acts as a sink for 0.02Gt/year. Overall the north Indian oceans contribute to 20% of total CO_2 emitted to atmosphere in South Asian region. 'Biological pump' in major part of world ocean is expected to change by CO_2 injection. Evidence has shown strengthening of plants can take place in low latitude regions especially equatorial Northern Indian Ocean and equatorial Ocean.

Second IPCC report however, noted that fertilization of Southern Ocean was not viable. It is believed that Southern Equatorial and North Pacific Ocean contain huge amount of unused micro-nutrients on their surface, but there is retardation of productivity as a result of biological pump. A few experiments are being planned to study iron fertilization effect in southern parts of the ocean. *Planktos* in USA[13] has proposed an iron fertilization for phytoplankton marine boom Weather Bird II project, where a 115 foot research vessel will transport and dissolve iron into Galapagos and South Pacific ocean over an area equivalent to 2.47 million acres of land to accelerate absorption of CO_2. An ocean fertilization experiment with 20 tonne of iron over an area of 1000 km in the South Atlantic Ocean is also being proposed by National Institute of Oceanography scientists in India.[14] Efficacy of this project is yet to be tested.

5.2 Regulations for Ocean Sequestration

Legal aspects of marine sequestration have been dealt in the number of regulations made from time to time.[15] Ocean legal framework applicable to CO_2 storage was established before it become an option to climate change mitigation. UN Group of Experts on Scientific Aspects of Marine Environment Protection (GEASAMEP) however suggested that laws may be required to be changed, it CO_2 marine sequestration is to be adopted. If

Ocean acidification occurs from CO_2 it will be classified as Hazardous waste but, hazardous waste Basel Convention may not be applicable for its Transboundary movement and ocean disposal. These laws are being reexamined for the purpose of regulating of CO_2 storage in Regional seas. The various marine regulations and International marine environment protection Laws from 1972 onwards are summarized in Table 2.

Table 2: International Regulations about Marine CO_2 sequestration

The Regulation	*Year*	*Remarks*
International Marine Environnent Protection Act	1972	Regulate dumping of waste and other matter and prohibits disposal of CO_2 into marine waters.
London Convention – Preservation of marine pollution by Dumping of wastes and other matter.	1972	Provide most extensive framework for all areas in marine environment protections and preservation.
UNCLOS United National Convention on Laws of Sea in force since 1994	1982	Acceptability of ocean disposal of industrial waste under international legal conventions
Basel Convention on Control of Tran boundary movement of hazardous waste.	1989	Preserved marine environments from Tran boundary movement of hazardous industrial waste
OSPAR Convention – Protection of Marine Environment of North – East Atlantic (Established by 15 European States)	1992	Most comprehensive frame work for North East Atlantic Ocean allows complex framework on CO_2 injection into geological structures in the subsoil in a manner that is unlikely to escape. Does not distinguish between storage in water column and off shore.
The Regulation	*Year*	*Remarks*

(Contd.)

(Contd. Table 2)

The Regulation	*Year*	*Remarks*
London Protocol Amendment	1996 Entered into forces in 2006	Prevent pollution caused by dumping or incineration of industrial waste. Provides for injection of CO_2 by pipelines form a land based source to a sub sea bed repository of oil.
OSPAR Convention - North East Atlantic	2002	Agree on placing of CO_2 is sea in consistency with London convention.
Amendments	2006	OSPAR committee recognized that ocean acidification and other effects of marine environment should be monitored in the context of CO_2 storage activities. Regulations about the impurity level in CO_2 injected are to be developed.
	2007	Adopted a decision to legally rule out placement of CO_2 into water column of the sea bed

Note: Definition of Territorial sea constitutes band of ocean stretching up to 12 miles. The Economic Zone..(EEZ) extends from Territorial sea up to 200 miles from the countries coast. Beyond this are High seas open to all States.

Separate conventions exist on protection of the marine environment of different seas. Convention for the Protection of the Marine Environment and Coastal Areas of South Pacific (LIMA, 1981) does not explicitly contain a prohibition on damping of waste in the sea bed or sub soil according to international standards. The Noumea Convention for Protection of the Natural Resources and Environment of the South Pacific Region as SPREP Convention came into force in 1990. The Convention covers the South Pacific Region to be 200 nautical mile zone established in accordance with International Law of the South Pacific Island Countries and those areas of high seas which are enclosed from all sides by 200 nautical mile zone of those countries. The Convention along the lines of UNCLOS required parties to prevent, reduce and control pollution of the ocean area from land based sources and from seabed activities. CO_2 is unlikely to be considered for marine storage under the Convention in its present form.

6. Conclusions

Threat of global warming and consequent climate change has given rise to most daunting challenge of 21st century. Carbon capture and storage provide technological approach for mitigation of CO_2 from the atmosphere. Most technologies for carbon capture and sequestions are in development stage and collaborative research program have began for their large-scale demonstration. The South Asia and South East Asia region is vulnerable to sea level rise and other climate change impacts. A summary of technologies for CO_2 sequestration and priority areas of research has been presented.

India has been taking proactive measures to initiate international collaborations in CO_2 sequestration technology and is taking part in the underground fixation feasibility studies in Deccan Continental flood basalt province. Plans to explore research possibilities of CO_2 capture and materials development of absorbents as well as enhanced CO_2 sequestration possibilities in crop and vegetation has been initiated. Scientific efforts being made worldwide to study ocean sink of CO_2 are also covered and a summary of regulatory decisions is presented in the context of Marine sequestration.

References

1. B. Ramachandra Rao and A. Sharma, Ed. Man and the Ocean, Selected papers presented at the 70th session of Indian Science Congress Association at Tirupati, Jan 3-8, 1983.

2. Malti Goel and H. N. Srivastava; Monsoon Trough Boundary Layer Experiment (MONTBLEX), Bulletin American Meteorological Society, (USA), 71, No.11, pp. 1584-1600, 1990.

3. B.S. Murthy, S.S. Parasnis and M. Ek, Interaction of land surface with atmospheric boundary layer: case studies from LASPEX, Current Science, 86, pp. 1128-33, 2004.

4. Malti Goel and Alok Kumar, Eds., Carbon Capture and Storage Technology: R& D Initiatives in India, Joint Publication of Department of Science and Technology and Ministry of Power, Government of India, pp. 167, 2007.

5. Malti Goel, Carbon Capture and Storage, Energy Future and Sustainable Development: Indian Perspective, in CARBON CAPTURE AND STORAGE R&D Technologies for Sustainable Energy Future, Eds Malti Goel, B. Kumar & S. N. Charan, p. 3-14, 2008.

6. D. M Kale, Research Issues in EOR using CO_2, in Carbon Capture and Storage Technology: R& D Initiatives in India, Eds. Malti Goel and Alok Kumar, Joint Publication of Department of Science and Technology and Ministry of Power, pp. 129-136, 2007.

7. R.C. Burruss, S.T. Brennen, 2003, Geological Sequestration of Carbon Dioxide – An Energy Resource Perspective, U.S. Geological Survey, FS-26-03.

8. Malti Goel, S.N. Charan and A. K. Bhandari, CO_2 Sequestration: Recent Indian Research, INSA - IUGS Indian Report 2004-2007.

9. Marchitti, C, Conceived the piping CO_2 to deeper depths of Atlantics, Climate change, 1, 59, 1977.

10. VVS Sarma, Influence of Carbon Sequestration on biogeochemistry of carbon in the ocean, Organized by

Department of Physical and Nuclear Chemistry and Chemical Oceanography, Andhra University, Sept. 21-22, 2007.

11. C. Prentici et.al. Climate Change 2001: The Scientific Basis, Contribution of WG I to Third Assessment Report IPCC, J.T. Houghte et.al Ed. Cambridge Univ. Press. NY 2001, pp. 183-237.

12. M.D. Kumar, Bio Geo Climate of North Indian Ocean, 1GBP-WCRP-SCOPE.

13. www.treehugger.com/files/2006/02/planktos.php.

14. Satya Prakash, R. Ramesh, M. S. Sheshshayee, Rahul Mohan and M. Sudhakar, Effect of Iron enrichment on nitrogen assimilation by marine plankton in the Southern Ocean, Current Science, 93, 2007.

15. Legal Aspects of Storing CO_2: Update and recommendations, OECD/IEA 2007.

6

Climate Change, Environmental Degradation and Societal Challenges

Col. P.K. Gautam

Introduction

Though South Asia may be politically divided into several sovereign countries, it is geographically the land, air and ocean mass of the Indian subcontinent. Ecological ecosystems overlap political boundaries. So do birds, wildlife, sea species, the weather and climate. The good health of an ecosystem directly relates to the economic potential of a country. Unfortunately ecosystems are getting degraded and both renewable and non renewable resource scarcities add to the problem of sustainable economic growth. Then there is the looming threat of climate change.

Common ecosystems and climate drivers that are shared by more than one country are the Himalayas and the big river system including ground water, Indian Ocean and coastal regions. Monsoons (except in the case of Afghanistan) including winter rains of the Westerly disturbances related to Central Asia provide the annual water budget. With Pakistan in particular India shares the Indus basin and the Thar desert. With Nepal India shares various river systems. With Bangladesh the Bay of Bengal delta with its mangroves is an important ecosystem. In the north east, jungle terrain of the region merges naturally with that of Myanmar and Yunnan province of China. With China, besides the water tower or the Himalayas common pastures such as the Changthang overlap Tibet and Ladakh. Unfortunately all ecosystems are under pressure brought about by greater need for resource use due to population and economic growth and the adverse impact of climate change.

National Circumstances

In human population and other human development index all countries of South Asia are medium development countries barring Afghanistan for which no data exists as yet. What is apparent is that we share poverty, we need to grow economically with sufficient rapidity to end poverty , we are energy deficient and we have an unsustainable population problem at hand which includes food security.

What is common is all nations are agrarian economies in terms of the bulk of the population deriving livelihood from agriculture and allied activities. Even with optimum growth of service sector or industrialisation it is unlikely that we can expect bulk of our population to be deriving their livelihood from other activities divorced from mother earth and mother nature.

Sufficient scientific evidence now exists that climate change is upon us. South Asian experience at the grass root level of farmers, pastoralists, mountain folks, fishermen and forest dwellers also shows a time series shift in ecology and weather for the worst. At the level of scientists and policy makers, The Stern Committee Report (*The Economics of Climate Change*) and the Fourth Assessment Report of the Intergovernmental Panel of Climate Change has shown that what we are experiencing is the tip of the iceberg.

Some Common Ecological Challenges

Climate Change. Though we are not part of the problem, we are the most vulnerable to the adverse impact of global warming as a result of climate change. Our adaptation is insufficient due to vulnerabilities such as being located in geographically vulnerable zones, poverty and coping capacity. Areas of common concerns are:

(a) *Sea level rise.* Submergence of small island nations such as Maldives, sea intrusion into Bangladesh and many parts of coastal India and Sri Lanka leading to ecological refugees (mostly from Bangladesh into India) is an emerging challenge. Population and infrastructure in coastal cities and towns is vulnerable to extreme events. Impact of sea level rise has already stirred the imagination of even novelists, as portrayed by Amitav Ghosh in his novel *The Angry Tide*. Though mitigation is being talked about in global negotiations, it is a losing battle of adaptation to sea level rise as it relates to developing countries. The small islands of the delta like Ghoramara in West Bengal are already submerged. A Jadavpur University study found that in the last 30 years, nearly 259 km of Sunderbans have vanished. Similarly in coastal Orissa, the Satabhaya region of Kendrapara district is being swept away by frequent tidal waves and sea intrusion up to 2.5 km inwards. This submergence in developing countries can be seen as a helpless situation in a losing battle against adaptation. This needs to be compared as a contrast to the successful adaptation being undertaken by the industrialised world. UK's annual flood and coastal defence (2004-2005) budget is US $ 1200 million, compare to which the international funding of $ 26 million for Least Developed Country Fund and Special Climate Change Fund is a joke. The conclusion is that there is a need for equality and justice in tackling the impacts of climate change. Mitigation must not be shirked away or palmed off to developing countries by the developed countries.

(b) *Melting of Glaciers and Impact on Water Resources.* Abnormal retreat of Himalayan glaciers and changed precipitation would cause havoc in the Indus basin and Indo Gangetic plains including the Brahmaputra valley. Over 600 million people are at risk. Initially floods followed by drought is forecast. Crisis over surface water sharing between India, Pakistan and Bangladesh is inevitable. Reduction in winter rain which is expected to be about 20 per cent will also impact on river flows between Afghanistan and Pakistan adding to tensions.

(c) *Agriculture.* Loss in productivity is forecast. Increase in frequencies of floods and droughts would lead to food insecurity. This will be in opposite direction to the need for more food grains as more food, fodder and feed is required for the rising population.

(d) *Forests.* Forests are important for biodiversity, biomass supply, watersheds and livelihood of forest dependent communities. Climate change is predicted to alter biome types, cause forest dieback and lead to loss of biodiversity which is already reducing due to habitat fragmentation.

(e) *Impact on Infrastructure.* Temperature and rainfall increase combined with sea level change and extreme events would demand risk assessment and vulnerability studies of coastal infrastructure.

(f) *Human Health*. Increase in vector borne diseases, health impacts related to extreme weather events and health effects due to food insecurity is forecast.

Death of the Oceans

In Dead Water. As most humans live mostly on land, the health of the ocean is taken for granted. Though the public at large is aware of abnormal retreat of Himalayan snow and glaciers - best demonstrated by the disappearance and shrinking of the snow lingam at Amarnath in Jammu and Kashmir in recent years – a greater disaster event is preparing to visit us. A United Nations study *In Dead Water* shows that Indian Ocean, the Persian Gulf and the Middle East of being adversely impacted because of global warming and rising marine pollution. It predicts 80 coral reefs with annual bleaching and death by 2080.

Carbon Sinks. V Rajamani Professor of Geology at School of Environmental Science at JNU has started a debate (*The Hindu*, 26 August 2005 and meeting report as given in *Current Science* Vol. 90. No.1, 10 January 2006, pp.12-13) on the ecological consequences of inter-linking of rivers on monsoons/ Bay of Bengal. There are some studies also on the relationship with carbon sink. Due to drop in water supply in the deltas/ ocean marine life would be deprived of nutrients. Also reduced runoff could effect indirectly phytoplankton population, affecting thus uptake of carbon/ the carbon cycle etc. Overall the research and knowledge on ocean as a carbon sink also needs to be deliberated as carbon sequestration may well be done in the oceans. One thing is now very clear is that extra carbon dioxide in the air is not only leading to climate change, but also making the ocean more acidic. Acidity which is a chemical reaction impacts adversely on the survival of marine life which in turn adversely affects the drawal of carbon dioxide from the atmosphere.

Shipping and Emissions. Emission reduction from ships and air transport is not included in any treaty as yet. Ocean traffic is increasing in relationship with worldwide trade and economic growth. From a share of just 1.8 per cent of global emission a few years ago, it has now been estimated that it constitutes 4.5 per cent of world emissions and by 2020 contribute six per cent of the global emissions. Thus oceans and its human users have a direct role to play in climate change.

Hurricanes/Cyclones. Evidence is mounting on global warming and the increase in intensity and frequency of cyclones. Its impact on shipping and livelihood needs to be appreciated as both observation and theory suggests that hurricanes are becoming more intense as the earth warms. A report in *Scientific American* showed that about eight per cent of the rain that hurricane Katrina dumped onto New Orleans in 2005, can be attributed to global warming.

Other Common Problems

If we set aside impacts due to climate change, the problems do not disappear. The India subcontinent shares a number of common ecological challenges. They existed even before climate change entered the political agenda. Some problems are:

(a) Agricultural soil degradation due to salinity, water logging, chemicalisation and erosion by water and wind.

(b) Deforestation including forest fires (problems with jhum cultivation as in Bhutan, India and Myanmar, and smoke from Indonesian fires engulfing the entire region).

(c) Desertification as in China.

(d) Sea water intrusion in deltas due to reduced fresh water discharge from rivers as in Indus and Cauvery delta as examples.

(e) Water stress due to demand and excess ground water extraction in Indus plains.

(f) Arsenic and fluoride in ground water effecting human health (Bangladesh and India).

(g) Coastal buffers such as mangroves and coral reefs degraded in all coastal countries.

(h) Intrusive species destroying biodiversity.

(i) Indoor air pollution and respiratory problems due to biomass based cooking.

(j) Atmospheric Brown Cloud and acid rain.

Disaster Events. The combination of environmental pressures and extreme events related to climate change point to more risks in the future. Some disasters are natural, but human interference with nature makes their impact much worst. The disaster events will have phenomenal impact on the economies. Unplanned urbanisation in floodplains, and drainage congestion due to natural buffers and wetlands being put under concrete or reclaimed for urban sprawls will be force multipliers of disasters.

Geopolitics of Climate Change

Artic Meltdown. The meltdown of the Artic has geopolitical implications. As ice melts, new routes as the northwest passage during summer for ships may open up, thus changing and shortening sea routes. But the most vital ecological issues are the rush for oil and the way the countries are claiming the region as an extension of the continental shelf. This may lead to militarization. What should be India's position ? Does it support it as being declared as a Arctic nuclear free zone as some scholars have alluded to by the negotiation of a treaty as in Antarctica? This issue needs to seen in the light of United

Nations Laws of the Sea and its ambiguities (of which the US is not a member). What is proper is that the the Arctic should be treated as a common heritage of mankind, rather than it being militarised which will further lead to its ecological collapse. The Antarctic Treaty System recognises the region of south pole as an area solely intended for scientific cooperation, keeps territorial claims in abeyance, and forbids all military activity in the larger interest of humanity. A similar international consensus would constitute the most appropriate response as suggested by the Worldwide Fund for Nature.

The Solutions

South Asia Association of Regional Cooperation(SAARC). Year 2007 was declared 'Green Year' in 14th SAARC summit. However though it focuses on disasters, it needs to shift to climate proofing by way of more water storage, more protection of coastal areas, higher capacity for flood mitigation, higher environmentally friendly building designs and global cooperation. SAARC countries now need to initiate dialogues on the climate related adverse impacts specially on the shared river systems. There is a need to have a common understanding of the natural events and work out ways cooperative means of overcoming the impact. We should not let disasters overtake us by being diplomatically proactive.

Conceptual and Educational Level. We need to treat climate change and environmental challenge as an opportunity. While the business class spurred by scientific research in appropriate technologies is hopefully bound to come to the planet's rescue, we need to focus on the social and behavioural changes that need to brought about. This is presently the toughest part. Ken Booth has pointed out that a desirable environmental road map should follow sustainable development principles, it ought to value nature, it needs to ensure fair share, it needs to have the polluter pays principle and should have a precautionary approach.[1] He further points out that differences occur between those environmentalists who believe societies can make progress relatively painlessly in capitalist global economy (as exposed by Al Gore and Jonathon Porritt) and those who argue otherwise and bid to enforce unpopular restrictions on well established lifestyle of rich (such as George Monbiot author of *Heat* and a regular writer in *The Guardian Weekly*). In South Asia we have the shining though now extinct or rare example of Gandhian frugalism and simplicity. Rather we can say that demand needs to be controlled rather than wants fulfilled unchecked by market forces. However this pathway would clash head on with the economic growth model based on competition and consumption. While one needs to balance between ecology and economic growth, what is important is to understand that we need to be eco-literate. Physicist Frtizof Capra explains that that we need to understand that that one species' waste is another species' food; that matter cycles through the web of life; that the energy driving the ecological cycle flows from the sun; the diversity assures resilience; that life, from its beginning more than three billion years ago, did not take over the planet by combat but by networking.

Public Participation. So far mass public participation in debating the issues of environmental degradation and climate change is conspicuous by its absence. Urban India in general appear to be the least concerned with it. Barring some committed social activists and Non Governmental Organisations (NGOs) on ecology, there are either sporadic seminars or discussions for the concerned intellectuals - which is like preaching to the initiated. Often adjustment of the common man to newer gadgets such as mobile phones, ipods and internet is given as example of the behavioural cum social changes that will be brought about by climate change and ecological destruction. But this does not appear to be the case. Strong social forces have redefined a consumer culture. Sociologist

Ashish Nandy while referring to the intellectual class has pointed out that previously there were people like writers, artists, teachers who were poor/austere. There was some vague awareness, amidst all hypocritical glorification of poverty, that social status did not have much to do with money. There was some space for some degree of austerity, voluntary poverty. This gave some dignity to those left behind in the rat race of life. Now we have rubbished this concept of poverty and collapsed it with destitution- which is undignified. Nandy further notes that the concept of poverty is expanding in such ways that many millions living comfortable lives have begun to think themselves as poor.[1] What is meant is that besides rising expectations there is also more demand for material comfort. As joint families break up and become nuclear or single, more space is required for the splintered members. Earlier a family car was enough. Now each member needs to own one. With these business as usual trends by the urbanites, lifestyle behavioural changes appear to be mere rhetorical arguments. Car owners of luxury cars often criticise the forthcoming induction of the small car "nano" as a recipe for disaster on roads due to congestion and pollution- but they themselves are not ready to walk or cycle to work or use a public transport or a best shed their sports utility vehicle.

We are at a crucial cross road in human history. One way is to invite ecological disasters and learn from it for a better future. This is the stupid way. A better way is to use or minds in a cooperative manner and work out social, technological and economic solutions so that we let not the climate impact on us adversely.

References

1. *The Hindustan Times* (New Delhi), April 12, 2007.

2. *The Hindustan Times* (New Delhi), May 18, 2007.

3. *Human Development Report* 2007/2008, p. 25.

4. As reported in *The Hindustan Times* (New Delhi), February 23, 2008.

5. Kevin E. Trenberth, Warming Oceans, Stronger Hurricanes, *Scientific American* India, July 2007, pp. 26-33.

6. Ramesh Thakur, " For an Arctic nuclear- free zone", The Hindu (New Delhi), September 21, 2007.

7. Editorial, *The Hindu* (New Delhi), September, 17, 2007,

8. Rieta Rahman, "Kyoto and South Asian Environment", *South Asia Journal*, October – December 2007, pp. 92-98.

9. Ken Booth, *Theory of World Security*, Cambridge University Press, 2007, p. 412.

10. Fritzof Capra, *The Hidden Connections*, New Delhi HarperCollins India, 2003, pp. 201-02.

11. Ashish Nandy, *Talking India: in Conversation with Ramin Jahanbegloo*, New Delhi: Oxford University Press, 2006, pp. 133-134.

7

Climate Change — Impact on Economic Development

Shyam Saran

There are two ways in which climate change could impact on our economic development.

Firstly, there will be additional costs due to the unavoidable need to adapt to climate change that has already taken place, and will inevitably take place, as a result of the cumulative and continuing accumulation of greenhouse gases (GHG) in the atmosphere. The economy will need to be made more resilient to face unpredictable weather patterns. It will have to evolve more thermal resistant crops and botanical strains. There may be severe water stress both for agriculture and human consumption. There may be health consequences due to the spread of viral and bacterial strains in higher temperatures and more humid conditions. The poorer the country, the greater the risks and, therefore, greater the costs of adaptation. Even currently, it is estimated that India is already spending about

2.5% of its GDP on adaptation and this percentage is likely to go up significantly in the coming years.

Secondly, there will be costs to be incurred if concerns over climate change compel a significant and accelerated shift in the energy mix of the country, away from the current reliance on fossil fuels to non-fossil fuels. The patterns of economic activity, in agriculture, industry and sciences, will need to undergo a transition which will require investments in new technologies and equipment. The economic and social development of the country and the prospects of eradicating poverty within a generation, would be seriously and adversely impacted.

Let me also emphasize at this point that ability to adapt to climate change is also linked to the level of development. Richer and more advanced states are better equipped to cope with climate change than are poorer countries. Therefore, development is the best form of adaptation, even if development in a developing country results, in the foreseeable future, to an increase in its GHG emissions.

There is much talk in our civil society today of the urgency of India taking on commitments to reduce its GHG emissions. Some argue that it is developing countries like India which will be worst affected by climate change and hence should take the lead in significantly reducing its GHG emissions. Some of our friends, both in Europe and North America also argue that unless countries like India and China take on ambitious commitments to reduce their GHG emissions, their own efforts will not yield the results we all hope for in mitigating climate change. Let us examine these arguments in the light of science and on the basis of equity.

Climate Change is taking place not as a result of current GHG emissions but as a result of cumulative accumulation of greenhouse gases in the global atmospheric space, as a result of anthropogenic activity, mainly fossil-fuel based industrial activity over the past 200 years. True, current emissions are adding to this accumulation on an incremental basis. But looked at in a comprehensive perspective, the total stock of GHG in the atmosphere, in particular CO_2 emissions, is almost entirely the responsibility of developed, industrialized countries. Even if one were to take 1900 as the baseline and calculate cumulative emissions upto 2004, the U.S. share in global cumulative CO_2 emissions is 28.03%, of Russia 7.85%, of Germany 6.47%, of Japan 4.02%, of China 8.37 and of India, only 2.44%. If you use the per capita basis, which is the only truly equitable standard, then differences are even more stark. The U.S. leads with 1733.3 metric tonnes between 1900 and 2004, Canada is second with 1250.4, the UK with 1111.2, Germany with 985.4, Australia with 921.5 and Poland with 686.9 metric tonnes per capita. For China, the comparable figure is 87.3 and for India 39.2 metric tonnes. In fact, among 21 top countries by per capita cumulative emissions, India is at the bottom of the ladder. (These figures are based on the Carbon Dioxide Information Analysis Centre of the U.S. Department).

Against this backdrop, whatever be the stabilization targets for CO_2 emissions, that is considered essential for mitigating climate change, there will be a corresponding carbon space which will need to be equitably shared among different countries in the world. It is clear from the figures that I have placed before you that equity demands that CO_2 emissions, which constitute the most significant part of GHG emissions, must be drastically reduced by the developed, industrialized countries, even while those of developing countries like India will inevitably rise, at least in the foreseeable future. If developing countries like India must mitigate their CO_2 emissions, then the costs involved must be fully compensated for, by the developed countries, and the required technology transfer must take place in the nature of public goods. This is based on the simple principle: The Polluter Pays.

In fact, what I am saying here today, has already been accepted by consensus in the UN Framework Convention on Climate Change, adopted in 1992 at Rio de Janeiro. The Convention recognized the historical responsibility of the developed countries for the phenomenon of climate change taking place. It is for this reason that only developed countries were expected to take on emission reduction targets and not developing countries.

An argument is sometimes advanced that in 1992, the rapid growth of emerging economies like China, India, Brazil and others, had not been foreseen. The situation, today, is different and, therefore, requires new measures. In fact, the UNFCCC clearly anticipated that, as developing countries pursue their goals of economic and social development, their emissions would inevitably rise.

As for current emissions, again India has no reason whatsoever to be on the defensive. I refer to the Climate ScoreCards brought out by WWF and prepared by Ecofys, on the eve of the recently concluded Hokkaido G8 and G8 + G5 Summits. If you look at current emissions per capita in tonnes of CO_2 equivalent, the figures are revealing. Canada has 23 tons, Germany 12, Japan and the UK 11 each, Russia 15 and the United States 24 tons per capita. China, by contrast has 5.5 and India only 1.7 tons per capita. Even in the macro terms, India lags far behind the U.S., for example, accounting for only 4% of global emissions as against 20% for the U.S. And despite their commitments in the UNFCCC to reduce CO_2 emissions by 6.2% in 2012 compared to 1990, most developed countries have in fact recorded a significant rise instead. This is because fossil fuel consumption is rising everywhere. In the 1990-2005 period, US emissions have risen by 19%, Japan's by 22% and even the

EU 15 by 7.8%. If you consider incremental CO_2 emissions on a per capita basis in select countries during 1990-2005, then the U.S. leads with 3.2 tonnes, China follows with 2.3 tonnes, Japan with 1.7, the EU 15 by 0.7 and India with only 0.5 tonnes. Therefore, it is absurd to talk of India being a "free-loader" or being given a pass. The problem is elsewhere.

I have gone into some detail in this regard because we must not accept a global climate change regime which is inequitable and imposes unfair economic burden on us as a developing country. Climate Change will become a constraint on India's growth story if we do not work together with other developing countries to ensure a global regime that takes into account historical responsibility, recognizes the overriding imperative of economic and social development and the eradication of poverty and is based on the principle that each citizen of the globe has an equal entitlement to the global atmospheric space.

Despite history and equity being on our side, we have not avoided our responsibilities as enlightened global citizens. The Prime Minister has committed India to never exceeding, during our development process, the per capita emission of developed countries. This is a major contribution to the global effort to deal with climate change.

I have so far argued that our negotiating strategy must be directed towards an outcome that does not constrain India's growth prospects by limiting our energy choices through emissions reduction commitments. I have also demonstrated that adaptation costs to India, as a result of current and prospective climate change, will be increasingly significant. In multilateral negotiations, our effort is to ensure funding of adaptation costs through an Adaptation Fund. However, even if there were no climate change constraints on India, would our energy security still be assured? In the next part of my presentation, I will touch upon this aspect briefly.

According to Report of the Expert Committee on Integrated Energy Policy, India's energy requirements will rise dramatically from the current period to 2030-31. It assumes a population size of 1.4 billion against 1.1 billion currently. It is assumed that in order to meet our minimal economic and social developmental goals, in particular to eradicate poverty within a generation, our primary energy supply must grow 3 to 4 times current levels and electricity supply achieve 5 to 7 times current consumption. Power generation must rise from a current level of about 1,60,000 megawatts to about 8,00,000 megawatts in 2030-31. Meeting this level of energy requirement necessitates the pursuit of all available fuel options and forms of energy. After analyzing current production and consumption trends and likely import patterns, the Report envisages the following:

- The dependence on coal, which is currently 51% may decline to about 45% in 2030-31, but nearly 2/3rd of the coal may need to be imported.

- Oil forms about 36% of our current energy mix, but this may decline to about 24% in 2030-31, but the level of imports may rise from 70% at present to perhaps 90% in 2030-31.

- Natural gas currently meets about 8.8% of our energy needs and this is likely to go up to 23% in 2030-31 on the strength of both domestic and imported supplies.

- Hydropower constitutes a little over 2% of our total energy profile currently and is unlikely to change very much by the year 2030-31, despite there being over 150,000 MW of potential hydro-power potential in the country.

- Nuclear Energy contributes only 1.5% towards our total energy mix but is expected to go up to 6.3% in 2030-31, but only if nuclear commerce with the rest of the world can be resumed and if India is enabled to import both nuclear reactors and uranium fuel.

- On current indications, without a major effort on R&D, renewables may only constitute about 6% of our energy mix in 2030-31.

What does this analysis tell us? If the energy mix remains more or less the same over the next 25 years, the overwhelming proportion of our energy needs will be met by coal, oil and natural gas. If over 66% of our coal, 90% of our oil and nearly 60% of our natural gas requirements have to be imported, India's energy security is going to be severely undermined. If, as is now obvious, there will be a growing gap between demand and supply of oil, gas and coal over the next decades, we will be confronted with much higher costs of energy supply. This is already happening, most dramatically in the case of oil, but also significantly in the case of natural gas and coal. And, therefore, even if there were no climate change argument, considerations of energy security alone would suggest a medium to long-term strategy of moving away from reliance on fossil fuels to non-fossil fuels, from non-renewables to renewables and from conventional sources to non-conventional sources of energy. And this shift, would, as a beneficial consequence, bring about major climate change gains as well, both for India as well as for the world.

It is this realization of the inter-linkage between climate change and energy security which underlies the strategy of sustainable development outlined in India's National Action Plan on Climate Change, which was released by the Prime Minister on June 30 this year. The National Action Plan gives priority to Eight National Missions. In terms of my presentation today, I would refer to two, which directly impact on energy security. The first is the pride of place given to Solar Energy among the missions, as an inexhaustible and renewable source of energy. We have to be ambitious in pursuing this National Mission because if we are able to come up with applications which are convenient and cost-effective, then there would be no energy constraint on our growth. There would also not be any climate change constraint either.

The second mission relates to Energy Efficiency. India has done rather well in virtually halving the energy intensity of its growth since the 1970s, from over 0.30 kgoe (Kg of oil equivalent) per US$ of GDP (PPP) to 0.19 kgoe currently. This is equivalent to the OECD average. This can, however, be further improved by about 25%-30% bringing us closer to the 0.12 kgoe, which is currently the best in the world. In coal-based thermal power plants, efficiency can be increased from the current 30.5% to 39% or more. The implementation of the Energy Conservation Act of 2001 and the Building Code of 2005, would bring about significant improvements in energy efficiency in most high energy-intensive sectors. The Integrated Energy Policy has an interesting concept of a "Negawatt", comprising the saving of energy through conservation, which has higher value than a Megawatt of newly generated energy. This is because in terms of a unit of energy saved, the costs of generation and transmission are all avoided.

There is little doubt that with the civil nuclear energy initiative back on track after the recent vote in Parliament, we may look forward to a major expansion in nuclear power generation in the period upto 2030-31. Domestic availability of uranium can only fuel 10,000 MW of nuclear energy for about 40 years. But India has very substantial reserves of thorium which can sustain a much larger and more sustainable nuclear power production. However, for that to happen fertile thorium needs to be converted into fissile material. Dr. Homi J. Bhabha, the father of India's nuclear programme, envisaged a 3-stage process, consisting of the setting up of uranium-based Pressurized Heavy Water Reactors in the first stage, leading up to the establishment of Fast Breeder Reactors in the second stage, based on plutonium reprocessed from spent fuel of the first stage and finally, reactors based on Uranium 233-Thorium 232 cycle in the third stage. Here thorium would be used as blanket material in Fast Breeder Reactors to produce Uranium 233 to fuel the third stage. While we have reached a degree of maturity in the first stage, the prototype 500 MW capacity FBR is currently under construction to launch the second stage. The pace of development of nuclear power is constrained by the rate at which plutonium can be bred and thorium converted to fissile material. The larger the scale of the first stage, the sooner can the third stage attain criticality, since adequate amounts of plutonium could be bred from the spent fuel of the uranium based reactors. In the optimistic scenario, contained in the Integrated Energy Policy document, nuclear energy could constitute about 63,000 MW in 2030-31 and over 275,000 MW in 2050. This would effectively make India energy-independent for the long-term.

Therefore, the civil nuclear energy initiative, far from being a distraction from the 3-stage indigenous nuclear energy programme, would actually contribute to its earlier realization and on a much larger scale than originally envisaged, based on domestic supplies of uranium.

Conclusions

Firstly, we must negotiate to establish a global climate change regime that does not restrict our energy options, nor imposes on us the costs of adjustment to a low-carbon economy. We are not in a position to meet these costs of adjustment unless these are fully covered by transfers of finance and technology from developed countries.

Secondly, even while we ensure that we keep our energy options open, we must recognize that India's energy security necessitates a graduated shift from fossil fuels to non-fossil fuels, such as nuclear, and from non-renewables to renewables, such as solar energy and bio-mass. The high and rising costs of oil, gas and coal and our increasing dependence on their imports, necessitates such a shift, and the sooner we achieve this, the better it is. Our ability to do this will also be greatly enhanced through international collaboration which we welcome. India's strategy for achieving energy security will also become its answer to the challenge of climate change, with benefits both for India itself and for the world.

8

Impact on Human Health

S. Venkatesh

Addressing the Challenge

In September 2007, a report of the Intergovernmental Panel on Climate Change (IPCC) followed by a high-level meeting organized by the UN Secretary-General clearly highlighted two facts: (1) warming of the earth's climate is primarily attributable to human activities; and (2) climate change is already disproportionately affecting developing countries. These are very real reasons for concern. The temperature of our atmosphere is projected to warm at the rate of about 0.2°C per decade. With the warming of the earth's atmosphere, the seasons are changing, glaciers receding and sea levels rising. The frequency and severity of floods and droughts are notably increasing, thereby challenging water, food and energy security; and placing an enormous strain on health security. Issues pertaining to the impact of climate change, especially in developing countries, were discussed in the UN Climate Change Conference in Bali,

Indonesia from 3 to 14 December 2007. It is clear that climate change is not only an environmental concern but also has profound health, economic and social consequences. It impacts disproportionately on the poorest and most vulnerable populations, undermining the gains achieved from poverty alleviation initiatives and threatening to derail the progress towards the Millennium Development Goals, putting them beyond the reach of many countries including those in the South-East Asia Region.

There is now a clear scientific consensus that the earth is warming, with more extreme events, such as heat waves, floods and droughts taking place, and that human actions are the dominant cause of these changes. Global atmospheric concentrations of carbon dioxide, methane and nitrous oxide have increased markedly as a result of human activities since 1750, and now far exceed the pre-industrial values which have been determined from studies of ice cores that span many thousands of years. The increase in carbon dioxide concentration has been primarily due to fossil fuel use and land-use change, while that of methane and nitrous oxide has been primarily due to agriculture. The evidence of warming of the climate system is unequivocal, and is observable in increases in global average air and ocean temperatures, widespread melting of polar and glacial snow and ice and rising average sea levels. Climate change is projected to accelerate over the coming decades, with temperatures increasing at approximately 0.2 degree Celsius per decade. Even if concentrations of all greenhouse gases and aerosols had been kept constant at year 2000 levels, a further warming of about 0.1 degree Celsius per decade would be expected.

Anthropogenic warming and sea-level rise would continue for centuries due to timescales associated with climate processes and feedback even if greenhouse gas concentration were to be stabilized. Such warming will induce many changes in the global climate system during the twenty-first century that would very likely be larger than those observed during the twentieth century.

Health and related risks

Climate change poses heath risks that are global, diverse and slow or impossible to reverse. Observational evidence from all continents and most oceans shows that many natural systems are being affected by regional climate changes, particularly temperature increases. These and related effects are likely to impact the health status of millions of people. The climate change is not only an environmental issue but also a health and developmental catastrophe. As earth's surface warms up, it brings in diseases that would challenge national and international health security as never before. These include increases in vector-borne diseases such as dengue, malaria and chikungunya; waterborne diseases especially diarrhoeal diseases, and malnutrition and cardiorespiratory ailments. Increases in the occurance of natural disasters in the form of floods and drought will lead to unprecedented degree of human migration, making them vulnerable to various communicable diseases. There is now strong evidence that the climate change that has occurred so far is already causing health effects, for example by increasing the likelihood of a heat wave like the one in 2003, which killed over 35000 people in Europe, and by contributing to conditions for malaria epidemics in the tropical highlands of many countries. Climate change affects some of the most important disease burdens, particularly those related to childhood, such as malnutrition, diarrhoea and also malaria and viral fevers. It will also affect populations that are the most vulnerable, and those who live in specific physical environments such as lowlying islands, river delta plains, and seaside areas. Communities that are poor and are inadequately protected by health services would also be severely affected. The greatest concern is for regions with the highest burden of climate-sensitive disease, the weakest public health capacity to respond to the additional demands for health services, and where climate change is most likely to slow the socioeconomic development. Unless actions are taken to reduce and respond to this threat, the disease burdens will increase, with the possibility of severe shocks to global public health.

Some of the health impacts of climate change include increases in the following:

- Acute deaths due to sudden heat waves, floods and droughts
- Vector-borne diseases such as dengue, malaria and chikungunya
- Waterborne diseases especially diarrhoeal diseases
- Malnutrition and its associated effects on child growth and development
- Morbidity and mortality related to cardio-respiratory ailments.

For example, climate change is expected to increase the proportion of the population exposed to dengue from about 35% to 60% by the latter part of this century. Similarly, uncontrolled floods will contaminate the water we drink leading to diarrhoeal diseases, and migration of people due to floods and drought will make them vulnerable to various communicable diseases.

The case of the Southeast Asia Region is naturally of serious concern to us because we inhabit it. But the vulnerability is especially great because this Region is already disaster-prone. Lowlying regions such as Maldives and other Indian Ocean islands; Bangladesh, and the Indian coastline are at risk from sea-level rise. More intense and frequent weather events such as cyclones and floods would result in the loss of precious lives and income, and leave scores of families and communities bereaved. Sea-water intrusion would cause fresh-water contamination, which could result in outbreaks of diarrhoea and a severe impact on affected populations. Furthermore, the displaced populations would be subject to potential psychosocial conflicts of adjustment to new habitats and alien environments.

Populations living in high altitudes may also get affected by these changes. Already the lives of those living in the high altitudes of the Himalayas are being affected by the risks linked to flash floods from bursting glacial lakes, caused by melting glaciers. Glacial retreat in the Himalayas will also shrink freshwater supplies not only in situ, but also downstream. Warmer temperatures will impact on biodiversity and upset agro-ecosystems, and probably will induce outbreaks of dengue and malaria. Moreover, climate change will disrupt rainfall patterns and bring about droughts that in turn would increase food insecurity, malnutrition and conflicts over water, and also precipitate huge population displacements.

Mitigation and adaptation

According to WHO's Director General, Dr Margaret Chan, climate change "may turn out to be the single most ominous struggle" facing the health sector in coming years. Therefore, the health sector must be empowered to be part of crucial decision-making processes in the context of climate change and be fully prepared to participate in mitigating the negative health consequences. Climate change and its consequences on health should be used as a powerful argument for substantive action on climate change. Action to address climate change is possible on two fronts. The first priority is to reduce emissions of greenhouse gases. Changes in lifestyle and behaviour patterns will be needed to contribute to climate change mitigation across all sectors. Management practices can also have a positive role especially those aimed at achieving higher energy efficiency. Although mitigating the impacts of climate change is paramount, it is also important to adapt to its impacts through preparedness and protective action. A wide variety of national policies and instruments are available to governments to create incentives for reducing greenhouse gas emissions and to initiate adaptation.

Several countries in the SEA Region have already taken action in this regard. Most have set up national bodies to specifically study national vulnerability and assess the potential impacts from climate change. Some countries have established national committees that are in charge of developing action plans to reduce greenhouse gas emissions and select adaptive measures. The National Climate Change Committee, established in 1993 by the Royal Thai Government, the Interministerial Committee on Climate Change established in Sri Lanka and the recently created Indian Prime Minister's Council on Climate Change are a few examples. All countries in the Region have preparedness and response plans for early warning systems, based on forecasting information that can be used to meet the challenges posed by climate change. For example, Bhutan has developed a plan to carry out concrete actions in five priority areas: by instituting medical emergency response; strengthening programmes aimed at controlling vector-borne and waterborne diseases; setting up an early warning sub-system by coordinating disease surveillance and climate-monitoring activities; reducing the risks of vector-borne and waterborne diseases by engaging and empowering local communities and through advocacy and awareness-raising measures in general.

While climate change is an issue that cuts across many sectors and demands a multisectoral approach, globally and at the country level, the health sector has a crucial leadership role. In order to reach for solutions that are beyond the health sector and create concern among such sectors, the health sector must articulate and share with other relevant sectors information on the potentially devastating health and developmental impact of climate change. In order to exercise health leadership responsibilities, we need to carry out health impact assessments at the regional, national and local levels on the extent and severity of communicable disease, as a consequence of climate change. Another area where the health sector can play a key role is instituting an early warning system which will further improve the preparedness of health departments to investigate and control outbreaks of vector-borne and waterborne diseases and strengthen risk reduction approaches through public awareness campaigns. To do this successfully, all relevant health programmes must strengthen their core capacity to be able to respond immediately and effectively.

Response of International Organizations

While critical evidence of the problem has been rapidly and effectively assembled, there is need to increasingly focus on raising awareness and supporting policy responses. This includes highlighting the health dimensions of climate change at major international meetings (annual conferences of the United Nations (UN) and ministerial meetings), WHO and other UN agencies to work on new initiatives and partnerships (Piloting health adaptation to Climate Change, with Bhutan as the pilot country in the SEA Region), and holding advocacy meetings and workshops. An attempt is also being made to urgently bring together the wealth of information on this topic that WHO, WMO, UNEP, UNDP and other partners have accumulated over the years into an organized and coherent whole.

The health sector, armed with this knowledge, would be in a strong position to advocate for the need to prioritize the health dimension of climate change and to increase opportunities for tapping the amounts of donor funding that would be forthcoming for climate change action in the future. This information will also help forge a thorough understanding on the part of national health and other leadership of link between climate change and health. Such understanding would hopefully ensure a sustained commitment to climate change-related efforts and needed strategic actions. Such a strategy might include six key elements: evidence-based actions; strengthening the key features of the public health system that are aimed at protection from climate change; guiding health development by promoting and advocating for energy-efficient national development options; advocating for individual behaviour change towards more environment-friendly choices; advocating with the private sector for investment and marketing approaches that would reduce environmental health risks; and monitoring and evaluating the effectiveness of climate change-related effects.

The Member countries in the SEA Region need to understand the linkages between climate change and health and to develop policy responses. The World Health Organization has been working in partnership with other agencies such as the World Meteorological Organization (WMO), the UNEP, and the UNDP to help the countries in the region. Three specific regional workshops have been held on this topic in the recent past. Most recently in 2007, a Workshop on Climate Change and Health in South-East and East Asian Countries was held in Kuala Lumpur, Malaysia, and was attended by representatives of countries from both South-East Asia and the Western Pacific regions; an Inter-regional Workshop on Human Health Impacts from Climate Variability and Climate Change in the Hindu Kush Himalaya Region was hosted by India in 2005; and the Synthesis Workshop on Climate Variability, Climate Change and Health in Small Island States was hosted by Maldives in 2003. In the words of Dr Samlee Plianbangchang, Regional Director, WHO Regional Office for South-East Asia, “Climate change is now high on the health agenda of WHO and we stand ready and committed to assist Member countries in planning strategies and building their capacities to mitigate the health impact of climate change in the SEA Region".

Way Forward

Ultimately, governments both in the developed as well as developing world must try to cut down on emissions creatively and invest in cleaner technologies, because therein lies the solution to preventing climate change and its catastrophic consequences on health. At the national level, high-level health policy directives need to be given to:

- recognize the primary role that the health sector must assume in developing and implementing national action plans for mitigation and adaptation to climate change;
- express, as a priority, national commitment to address the challenges posed to human health by climate change;
- provide clear directions to plan actions and investments at the national level to address the health effects of climate change and
- promote the wider objective of strengthening the preventive approach to public health.

9

Impact on Gangectic Delta

Kakoli Banerjee, Avijit Gangopadhyay and Abhijit Mitra

Introduction

Climate change is one of the most critical global challenges in the present *era*. Recent events like flood, hurricanes, tornadoes and forest fires have emphatically demonstrated our growing vulnerability to climate change. Climate change impacts encompasses several sectors like agriculture – further endangering food security, sea level rise and the accelerated erosion of coastal zones, increasing intensity of natural hazards, species extinction and the spread of vector borne diseases. The impact of climate change on the aquatic ecosystem is an interlinked event between the melting of polar ice or glaciers feeding the rivers and the alteration of salinity in the riverine and estuarine waters of the tropics, temperate and sub-temperate zones. Thus, climate change is not one story, but many parallel stories of several ecosystems - only sporadically connected so far without very concrete data base. We have attempted to interlink the source and the sink ecosystems in this paper to detect the foot print of climate change at the sink (aquatic system of Indian Sundarbans).

Our approach

The study of climate change in coastal and deltaic complex of Indian Sundarbans needs to be addressed through an integrated approach as the area is the final receptacle of all waters from the upstream zone, river discharges and run-off. The salinity of the aquatic phase is regulated by several physical, geographical, geological and even man-made factors. Some important events like evapo-transpiration, riverine discharge, siltation, discharges through barrages constructed in the upstream areas etc. alter the salinity at the river mouth. Hence, all these parameters need to be considered to analyse the variations in hydrological parameters. The geographical scenario of the western and eastern parts of Indian Sundarbans is different as the rivers in the western part are connected to the Himalayan glaciers, whereas the rivers in the eastern sectors are all tide fed. Hence, to achieve a realistic picture of climate change, we collected data on surface water temperature, salinity and dissolved oxygen after compartmentalizing the deltaic complex into east and west.

Results and Discussion

The surface water temperatures in both the sectors have shown a rising trend (Fig. 1), a signal of global warming in this part of the Indian sub-continent at the apex of Bay of Bengal.

The pulse of the phenomenon of temperature rise is visualized in two contradictory ways in both the sectors due to variation in geographical features. The rivers in the western sector of Indian Sundarbans (Hugli and Mooriganga), being the

Fig. 1. Increasing trend of surface water temperature (in °C) in Indian Sundarbans

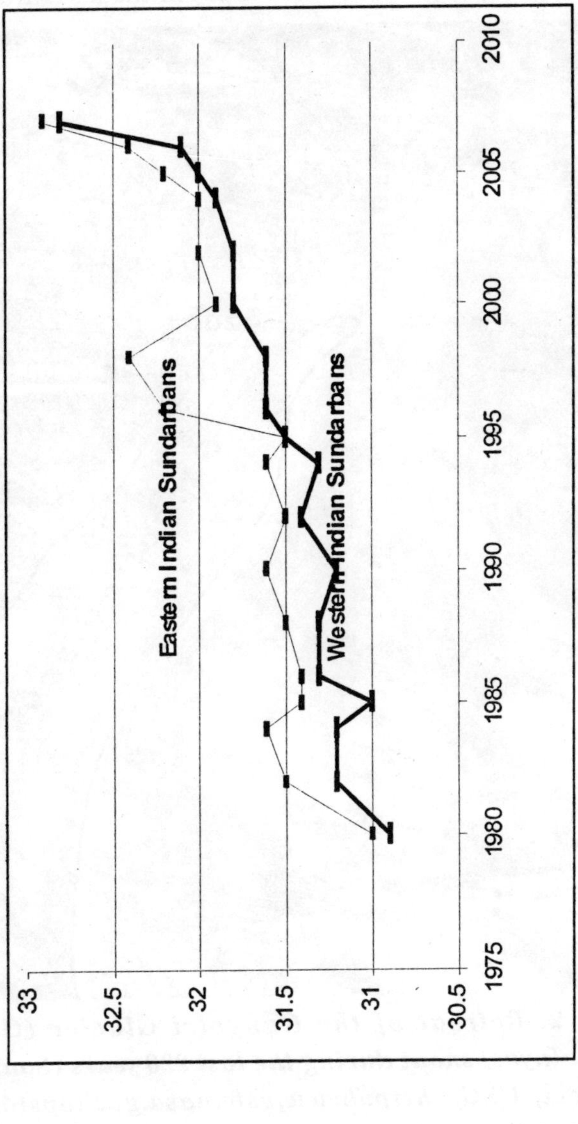

continuation of Ganges, receive the snow melt water of Himalaya. The impact of temperature rise is perhaps already upon the Himalayas. The 30.2 km-long Gangotri Glacier is receding rapidly: the rate of retreat in the last three decades

Fig. 2. ***Retreat of the Gangotri Glacier (Garhwal Himalayas) snout during the last 220 years (Source: Jeff Kargel, USG;, http://www.gsfc.nasa.gov/topstory/2002/20020530 glaciers.html)***

has been found to be more than three times the rate during the earlier 200 years or so. The average rate of recession has been evaluated by comparing the snout position on 1985 topo-sheet map and the 2001 satellite panchromatic satellite imagery and the results show the average recession for this period is about 23m/yr (Hasnain, 1999; 2000; 2002). The pictorial plot based on historical evidence and recent data on Gangotri Glacier retreat in a research by Jeff Kargel, Geologist of USGS also supports the increased rate of retreat of the Gangotri (Fig. 2).

The *ex situ* effect of Himalayan deglaciation is confirmed not only by lowering trend of salinity (Fig. 2), but also through gradual increase of dissolved oxygen (Fig. 3), in spite of rapid industrialization and urbanization in the western part of Indian Sundarbans receiving fresh water supply through Ganges conveyer belt system.

Such trends of lowering salinity and increasing dissolved oxygen level are absent in eastern sector, and here the foot print of global warming is felt through rising tidal amplitude, increased salinity (Fig. 3) and lower dissolved oxygen in the river mouths (Fig. 4) due to intrusion of saline water from the adjacent Bay of Bengal region.

The results of different foot prints of climate change may pose serious problems in livelihood and biodiversity spectrum of the area. Increased flooding in the adjoining cities and towns of western Sundarbans was already observed since the last few decades. The livelihood of the people in the eastern sector is under threat due to salinization of agricultural land, loss of commercially important fisheries in the catch basket, vanishing of fresh water loving Sundari (*Heritiera fomes*) trees etc. While deglaciation and subsequent change in hydrology has every possibility to hinder economic profile of the deltaic complex, the impacts are likely to be hardest felt by the people living

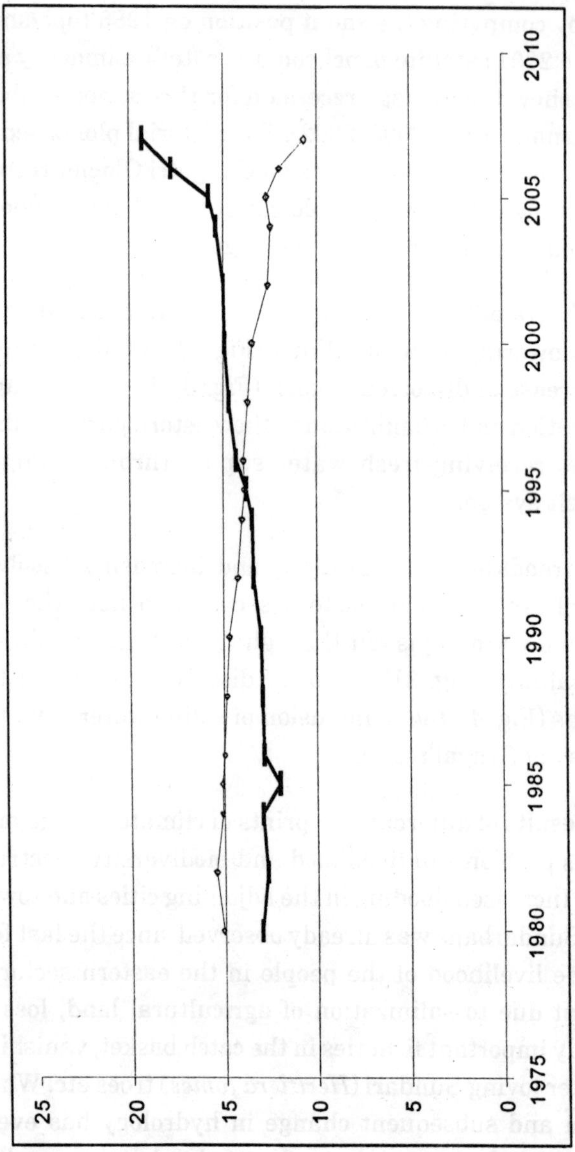

Fig. 3. Salinity (in parts per thousands) fluctuation due to global warming in eastern (rising trend) and western (decreasing trend) Indian Sundarbans

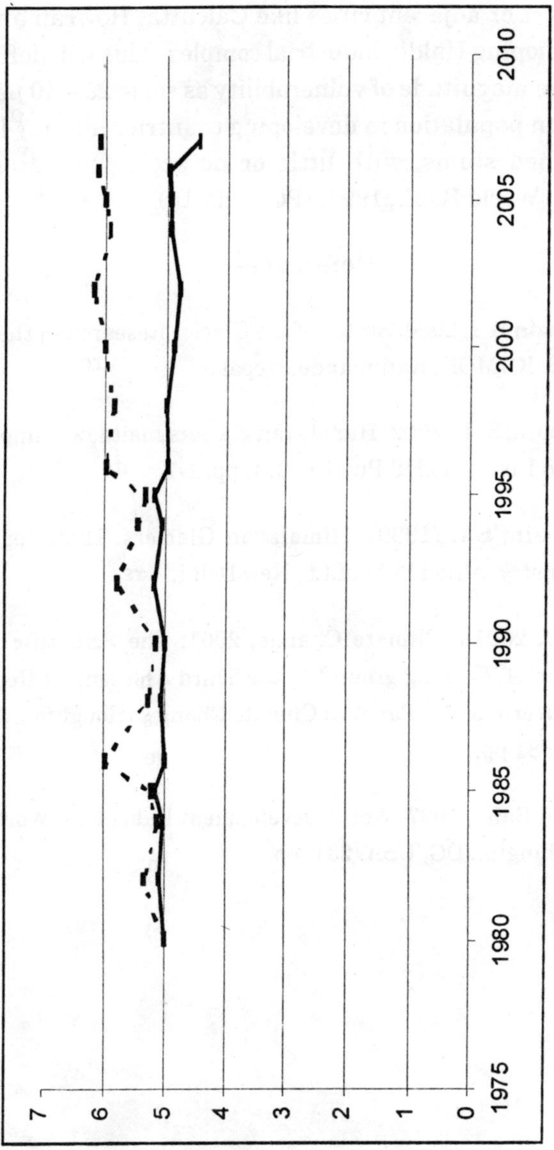

Fig. 4. Dissolved oxygen (in ppm) fluctuation due to global warming in eastern (decreasing trend) and western (increasing trend) Indian Sundarbans

below poverty line. Pressure on livelihoods may force poor landless island dwellers of the eastern sector to migrate in the western part or adjacent cities like Calcutta, Howrah and the newly developing Haldia industrial complex. This will definitely magnify the magnitude of vulnerability as some 25 – 40 percent of the urban population in developing countries already live in impoverished slums, with little or no access to water and sanitation (World Bank, 1997; IPCC, 2001b).

References

Hasnain S. I, 2000. Status of the Glacier Research in the HKH region.2000, ICIMOD, Kathmandu, Nepal.

Hasnain, S. I., 2002: Himalayan glaciers meltdown: impact on South Asian Rivers. IAHS Pub No. 274, pp. 1-7.

Hasnain, S.I. (1999). Himalayan Glaciers: Hydrology and Hydrochemistry, Allied Publ., Ltd., New Delhi.

IPCC, 2001b. Climate Change, 2001: The Scientific Basis. Contribution of Working group I to the Third Assessment Report of the Inter-governmental Panel on Climate Change. [Houghton, J.T., et al., (eds.)]. 881 pp.

World Bank, 1997. World Development Indicators. World Bank, Washington DC, USA. 231, pp.

10

Enhancing Adaptive Capacity to Climate Risks: An Experiment in the Semi-Arid Regions of India

Isaac Manuel. R.

Introduction

Around 30 years ago, the world faced a global food shortage and experts predicted that it would result in a catastrophic famine. However that danger was averted due to an intensive international research effort which enabled the development of high-yielding varieties of major food crops. This "Green Revolution," driven by knowledge and investment, was effective during that time because environmental conditions were favourable, soils were fertile and water was plentiful. Presently in the 21st century we face yet another more devastating environmental threat, namely global warming and climate change, which could cause irreversible damage to the ecosystems causing a great threat to the survival of mankind. Nowadays it is common everywhere around the globe that nearly every day we hear something about Global Warming or Climate Change. Intense storms, frequent floods, severe drought, species extinctions, melting glaciers, increases in the ranges of disease vectors and the list go on.

"Global Warming – What's All that about?"

Over the past 50 years, human activity has altered our ecosystems more rapidly and extensively than in any other comparable period in history, this was largely to meet the increased demand for food, clothing and shelter. Industrialization has lead to an increase in the concentration of greenhouse gases, disrupting the natural balance and in turn augmenting global warming. This increase in temperature in turn leads to a set of changes in the earth's climate or long-term weather patterns that varies from place to place. It may also lead to changes in the frequency and intensity of extreme weather events, though it is difficult to connect specific events to global warming.

Why is this, a concern?

Developing countries like ours tend to be more vulnerable to extreme climatic events and adverse impacts of a gradual climate change than industrialised countries, mainly because of higher reliance on climate sensitive sectors, such as agriculture, fisheries, forestry, water and other natural resources. Diversified agro climatic conditions, different cultures and ignorance about environmental changes due to lack of education and awareness paves way for limited capacities to anticipate and respond to climate change. Limited human, institutional and financial capacity of governance also acts as a set back to cope with these negative impacts of climate change.

Have we done anything to rectify the problem?

While scientists around the globe agree that climate change is already been occurring, the emerging impacts of the changes are yet to be realized by most farming communities in India. However, farming communities have faced adversity associated with climate variability and responded and adapted informally, in an ad hoc manner, for centuries. They have used their native wisdom as the basis for decision making regarding food security, human and animal health and natural resource management. Local coping strategies have come handy many a times. Nevertheless, the forces of globalization in operation today, coupled with the increase in climate related risks, have exacerbated and exposed the gaps in such local adaptation strategies.

What is Adaptation to climate change?

Adaptation is an adjustment process through which people reduce or moderate the adverse effects or potential damages due to climate change including climate variability and extremes. In simpler terms adaptation means taking action to reduce the adverse consequences of climate change, or to take advantage of any opportunities it may present.

Adaptation may be autonomous or planned. Autonomous adaptation tends to be what people and systems do, as impacts of climate change become apparent, for example, the migration of animal populations seeking cooler temperatures in response to climate change. Planned adaptation are measures taken to reduce potential anticipated risks of future climate change for example limiting or re-designing developments in areas that may be affected by future sea level rise. When planning adaptation responses, we can either take action now based on our existing knowledge of climate change and potential impacts, or we can respond to the change when it occurs. Specific decisions on whether we act now or wait will need to consider a number of issues including costs and benefits and the time it may take to respond to determine the most effective response. Benefits of early action may help reduce the need for and the cost of remedial action and may also provide some additional benefits in the interim.

Vulnerabilities of semi-arid areas of India

India's recent initial National Communication to the UNFCCC[2] confirmed that the livelihoods of the majority of the Indian population are threatened due to the impacts of climate change. Besides the coastal zone the arid and semi-arid areas

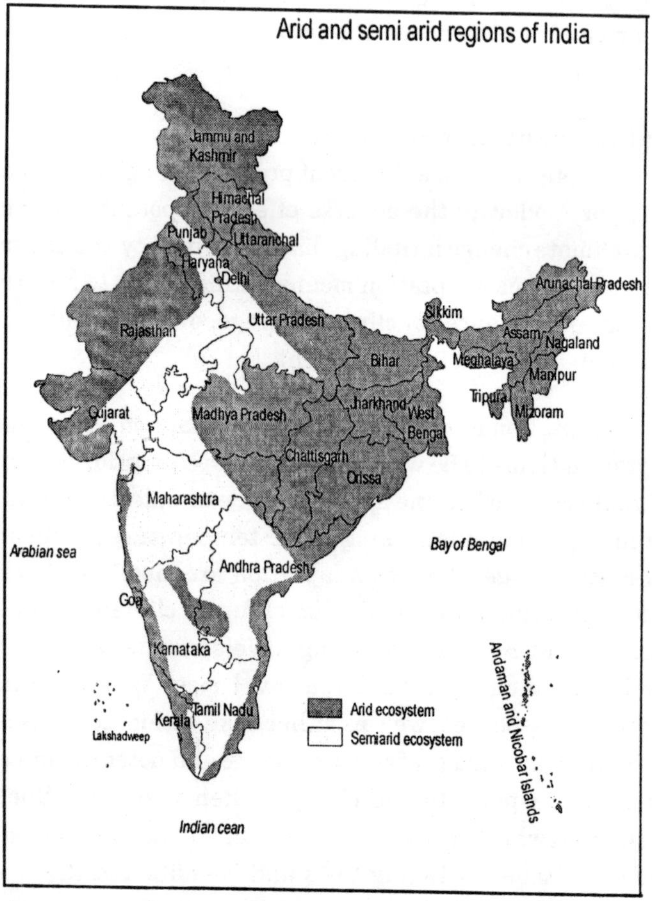

Fig. 1: Arid and semi-arid regions of India (Source: Velayutham et al., 1999)[2].

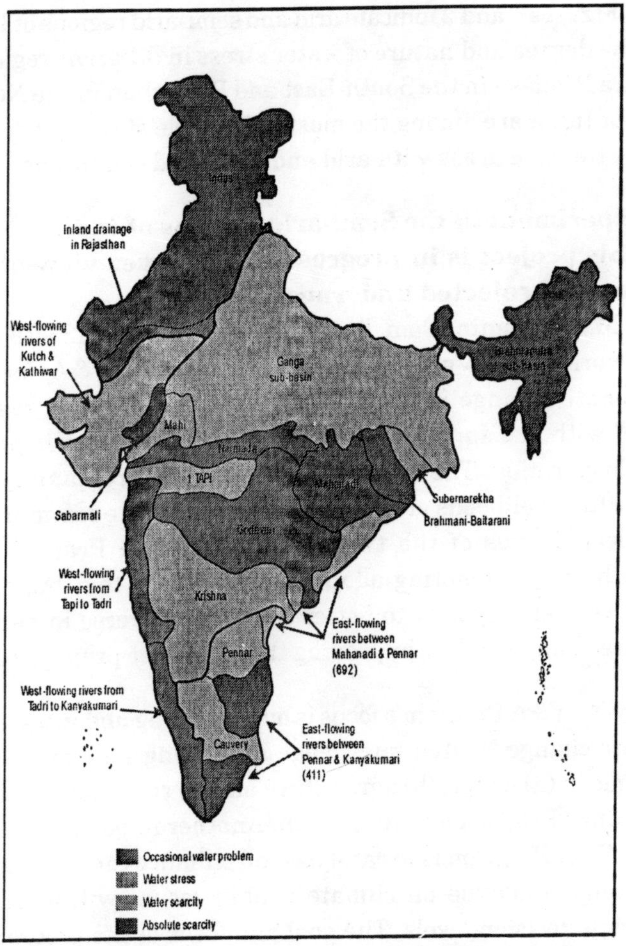

(*Source*: TERI GIS Database, 2001, based on Chitale, 1992)[v]

Fig. 2: Degree and nature of water stresses.

are considered to be among the most vulnerable regions in India. Arid and semi-arid areas in India experience an annual rainfall between 100 and 400 mm or 400 and 800 mm respectively with a very high coefficient of variation ranging from 40 to 70%. Low and erratic rainfall coupled with extreme temperatures and intense solar radiation results in frequent crop failures and considerably affects the agricultural economy in these regions[3].

Figures 1 and 2 indicate arid and semi-arid regions of India and the degree and nature of water stress in different regions.[4] Andhra Pradesh in the South-East and Rajasthan in the North-West of India are among the most vulnerable states because of their extensive areas with arid and semi-arid conditions.

An experiment in the Semi-arid regions of India (As this project is in progress only an over view of the works are projected underneath)

M.S. Swaminathan Research Foundation (MSSRF) has been working on "Vulnerability and Adaptation (V&A) aspects of Climate Change in the selected sites of Semi-Arid regions India" with the support from Swiss Agency for Development and Cooperation. The overall goal of the V&A Programme is to secure the livelihoods of rural poor and vulnerable communities in selected sites of the two states of Andhra Pradesh and Rajasthan by promoting adaptation measures that build and enhance their capacity to better cope with adverse impacts of climate change and by improving their disaster preparedness.

The V&A Program's focus is on optimising and integrating climate change related knowledge in existing service delivery systems in (a) water, (b) agriculture and (c) rural energy sector in Andhra Pradesh and Rajasthan. Another important feature of the V&A Program is to catalyse and enhance communication and policy dialogue on climate change issues within and at different decision levels. The goal and focus of the programme have been articulated through three specific objectives that integrate the local community, the state/district as well as the Indian national and international level. These objectives are:

- to build community level capacities with regard to best practices and technologies in the agriculture, water and energy sector,

- to optimise the service delivery system and services (extension services) at selected sites of Andhra Pradesh and Rajasthan, and to promote general awareness on climate issues and promote policy dialogue and advocacy at different levels both nationally and internationally.

To achieve these objectives the V&A Programme combines four different program components that help to define the way of implementing activities: research and analysis; natural resource management; capacity building and action learning; and local empowerment.

The research and analysis component covers the development of a local knowledge base on the effects of climatic hazards, on the availability of water and energy resources and on agriculture production systems. It aims at understanding the strategies and techniques that local communities have developed to cope with these climatic hazards. Activities like stocktaking and documentation of local practices, situation analyses and the analysis of biophysical vulnerabilities are included under this component. The research component also explores methodologies how to interlink the local knowledge and local coping mechanisms with the vulnerability assessment and adaptation to other ongoing national and international initiatives and projects.

A resource management component covers interventions that aim at developing sustainable management practices on a pilot scale. It includes the participatory development of techniques and technologies as well as social agreements for managing water, agricultural production and energy production, and distribution at community level. Activities like improving management skills, validating existing techniques (traditional, indigenous) and introducing technologies for resource management, transferring best practices, introducing tools for weather monitoring, and producing management guidelines and manuals for V&A are part of this component.

The component of capacity building and action learning aims at improving awareness and in-depth understanding on V&A at different levels. In addition, the programme will support actions that improve the use, effectiveness and efficiency of the service delivery system as a means to reduce vulnerability in rural areas. The programme will support activities like information, dissemination, training as well as the establishment of hubs and knowledge canters.

The local empowerment components described above will promote local empowerment through improved knowledge and negotiation skills at different levels. To do so the programme includes activities to improve the information flows, the functioning and creation of a platform for policy advocacy, the design and use of information systems and mainstreaming lessons learned into the national and international decision-making process. These components are complementary and promote the holistic understanding on which this V&A programme is based upon

So far under this project detailed analyses of the past and current climate scenarios were carried out in the project regions to underpin adaptation strategies. A variety of 'best practices' to handle climate related risks have been documented through a multi-stakeholder consultative process and direct field verification and work has been initiated to implement them with farmers' participation in the project areas. Besides taking stock of the informal adaptation innovations and providing synergies with relevant contemporary knowledge and technologies, the project has developed strong capacities at the local level on various aspects of weather predictions, risk management and crop advisories through 'Village Knowledge Center' (VKC). The V&A project has an inbuilt VKC component to disseminate timely weather related information and other information pertaining to the livelihood security of the local communities besides promoting general awareness on climate related issues. The insights gained during the 'Situational Analysis' phase of the project guide current project implementation efforts and provide a vision for sustainable application of climate information in agriculture and implementing suitable enhanced adaptation measures.

Milestones reached so far

Enhancing the adaptive capacity of the local communities

- Best coping practices identified, refined and introduced
- Fuel efficient good stoves introduced to conserve energy
- Efficient community water management practices introduced
- Weather based farming decisions inculcated
- Joint and community pasture land developed and maintained
- Improving goat breeding to enhance the adaptive capacity
- Various capacity building activities initiated to enhance coping
- Various awareness campaigns undertaken

Improve delivery system

- Current curricula for extension training examined and relevant course materials on climate variability and change are being developed.

Promote multilevel policy dialogues

- Possible policy avenues explored for up scaling effective coping mechanisms from grassroots level to national level
- A draft Climate Change Act is being prepared
- A new Vulnerability tool in the form of a calculator was developed to identify vulnerable regions to climate stress.

It is expected that within an evolving international climate change context, this V&A programme in India will make an important contribution to "Adaptation Science". The program's aim is to demonstrate and implement adaptation measures and coping mechanisms at the community level and at the same time to translate such action learning results to an international policy level. Experiences, lessons, mechanisms and techniques identified to enhance adaptation at the local levels through this project will be communicated for decision-making at the national/international level and shall be mainstreamed into the process of establishing India's Second National Communication to the UNFCCC.

References

1. Ministry of Environment and Forests, Government of India. 2004: India's Initial National Communication to the United Nations Framework Convention on Climate Change.

2. Rao, A.V., B.K. Garg and D.C. Ojha (An Annotated Bibliography). 2000. Desertification: An Indian Scenario. Jodhpur, Scientific, 284. ISBN 81-7233-230-0 of Water and Fodder.

3. TERI. 2001. Appraisal of initial steps towards a V&A framework for India with special focus on water resources, Final Report prepared for INFRAS, TERI Project Report No. 2001GW43, December.

4. Velayutham M., D K Mandal, C Mandal, and J. Sehgal. 1999. Agro-ecological Sub regions of India for Planning and Development. NBSS Publication 35. Nagpur: National Bureau of Soil Survey and Land Use Planning. 372.

5. Chitale M.A. 1992. Population and Water Resources of India, *In*: Gowarikar V *ed*. The Inevitable Billion Plus: Science, Population and Development. Pune: Unmesh Communications. 452.

11

Sustainable Solutions in Organic Farming

Juli and Vivek Cariappa

Introduction

In any agricultural operation, whether big or small, we are faced with two critical issues:

1. the need to feed humans and animals

2. the need to ensure that the biological life in the soil, air and water are sustained to ensure optimal photosynthetic activity taking place through plant organisms thereby utilising and diffusing solar energy.

As human beings who have reproduced more than any other animal species and effected major changes in our environment, we become responsible for interacting with these living networks in a manner where we can extract what we need without disturbing the generous margins that have been provided within the natural design. That today we have come to a point where we have obviously abused these margins to the point of having to question our ability to survive as a species on Earth is not a bad thing. Perhaps we are beginning to understand the finiteness of matter and that there are consequences to every action we take; the time for taking the Earth for granted is over.

Agriculture is an activity human beings have to do to survive. History is full of examples of how we have done this activity and we must have been getting something right to have survived so far. But the brutal fact is that "the changeover from hunter-gatherers to nomads to peasants took thousands of years, the change-over from peasants to farm-managers took only fifty years"…. In that fifty years there was a vast shrinking of diversity in the types of food grown. In 1984 an FAO study claimed that 90% of the human population consumed less than 40 animal and plant species and it is likely that in 2008 the situation is even worse.

The point is that the cultivation of this narrow food basket has been done in a way to suit industrial economics, over-using precious non-renewable fossil fuels, over-exploiting topsoil both physically, in terms of in-appropriate ploughing giving rise to soil and wind erosion as well as chemically, in terms of disturbing the balance and availability of atomic elements. Mono-cropping, massive and wasteful use of petroleum based agro-chemicals which require heavy irrigation, soil compaction due to the large machinery used and the lack of biomass returned to the soil to replace lost humus, gives an overall view of the huge loss of carbon directly and indirectly that has significantly contributed to global warming.

The effect of the cocktail of pesticides that have accompanied this model, needless to say, has deeply affected the biological checks and balances required for a resilient agriculture. Modern farming systems have done much to erode the very systems that for eons have played a significant role in modulating the balance of life, particularly microbial life in the soil and air, and this has impacted micro and macro climatic conditions on the Earth.

Organic farming on the other hand is dependent on renewing the living soil and re-creates through its inherent need for bio-mass, oxygen rich environments which incorporate carbon rather than release it, while absorbing moisture for slow release. By its very nature, this farming system imitates the forest ecology from which we all have evolved. This may indeed be what is required to meet the challenge of a climate change that is by estimation of the experts, a change far exceeding the normal expected changes in Earths' climate. We must view organic farming then, not only as a system of food production, but also in terms of its wider effects of increasing the carbon content of the soil, increasing the water holding capacity of the soil and sustaining a diversity of plants and animals to balance a closed circuit of energy on Earth: as a system that is well positioned to contribute significantly to resolving the dilemma of galloping climate change.

Why organic farming?

Organic farming, in whatever form or label it comes under, seeks ;

- To utilise natural resources within the given farm microcosm in such a way as to create an in-built principle of re-cycling, taking no more than it is capable of returning

- To optimise the energy ratio, so that the energy expended to produce an effect should result in a geometrically expanding spiral effect in a multi- dimensional way. This is with particular reference to photosynthesis and the conservation of water

- To maximise the integral life principle and seek to understand the balance between air, water and soil and all the linkages between them in the form of bacteria, fungi and plant life as well as the role of animals in perpetuating ecological niches

- To create the conditions for plant biodiversity to ensure the perpetuation of a reliable food supply throughout the food chain
- To maximise productivity while sustaining the ecology, without compromising on the nutritive and other qualities of the crop
- To ensure that the process of farming adds to the vitality and fertility of the soil rather than depreciating it and depleting its life carrying potential. The appreciation of the soil will lead to a nutritional balance and thus ensure a resilience in the health of all those creatures that depend upon it for life.

The human-soil relationship is largely concerned with the stewarding of plants. We need to fertilise and care for the soil because we have interfered with the natural cycle found in the wilderness. *Agri* is derived from the term ager which means field and *culture* is tillage of the field. Plants represent, to the creatures who depend upon them, the link between the sun, air and soil. The thousands of varieties of plants that exist on this Earth are an essential part of the alchemy of transforming the main constituents of life: carbon, oxygen, hydrogen and nitrogen, into food, first for themselves through photosynthesis, and then for creatures such as us. Plants are also a critical element in the process of *pedogenesis*, or soil building. It is now established that 95% of a plants nutrition comes from the air (atmosphere) i.e. its quantitative nutrition. The complex root systems of plants are designed to seek out its qualitative nutrition, the remaining 30 other elements, from the soil through water and soil fauna and flora with which they share symbiotic relationships.

elements from the atmospherecarbon, oxygen, hydrogen and nitrogen constitute 92% to 98% of a plants dry wt.**elements from the soil**2-5% of a plants dry weight 12 vital elements: 2 are non-constitutive-potassium and chlorine 10 are constitutive- phosphorous, boron, calcium, magnesium, Sulphur, iron, molybdenum, copper and zinc 18 secondary micro-nutrients:4 non constitutive- lithium, sodium, rubidium and cesium14 constitutive- fluorine, silicon, selenium, cobalt, iodine, strontium, barium, aluminium, vanadium, tin, nickel, chromium,beryllium and bromine

Table Courtesy Claude Bourguignon

The benefits of diversity in plants are :

(a) to cushion the Earths crust with differently designed plants to guarantee a capability of growing in combinations of diverse conditions to perpetuate photosynthesis

(b) to circulate and recycle the elements evenly above and below the soil which explains the variations in root-system design and purpose

(c) to play an important part in the aeration of the soil as well as be key factors in the carbon and nitrogen cycles

(d) to furnish balanced food for the process of humification and insure that the clay-humus (mineral-organic matter) complex is bound together dynamically.

(e) to regulate the temperature, moisture and atmospheric conditions around the living earth.

Obviously it is crucial for our own survival to protect the diversity of plant life rather than diminish it. One hundred years ago agriculture across the world was a celebration of variety. In

India we had thousands of varieties of rice and minor millets and at least 50 varieties of pulses. All regions of the world were

Imitating the forest ecology in organic farming.

rich in diversity of fruits and vegetables and grains adapted to regional variations in climate. As varied as the food was, so were the people culturally diverse and proud of the foods that were unique to their regions. Today's' industrial agriculture has led to large food stocks but quantity and productivity have not solved hunger (as the questions of distribution and accessibility have not been addressed), but instead have brought in their wake a rate of desertification that is one of the prime reasons for global warming. Not only are we losing at least 12 million hectares of forest annually but about 11 million hectares of arable land are lost due to desertification, toxification and conversion of cropland to non-agricultural use. Worse yet the loss in diversity and quality of food, has come hand in hand with plants and animals needing much more intervention than ever before and in the words of agrologist Claude Bourguignon, "we are now practising not agriculture, but pathological damage control".

To not only protect plant diversity but ensure the Earth is adequately green we have, as a first step, to use all of our knowledge in renewable energy to reclaim our vast agricultural fields through a far-reaching programme of land reform and agricultural policy reform in which composting is a core element. Fertility is never a given. Yet each bit of Earth is capable of being fertile. The only really significant change we make as a short-lived species is the condition we leave the space we occupy during our lives. This is determined by our consumption and production choices and so we return to the point that to eat we must farm and by our farming create fertile soil, fertile air and a green canopy on the Earth. We take inspiration literally and metaphorically from the forest, as a successful model of a micro-climatic bubble that will play its own miniscule part in modifying and controlling the climate that sustains GAIA.

Since we have de-nuded vast areas of the Earth for agriculture our first obligation is to undo the devastation of chemical agriculture which has seriously degraded the symbiotic humus-clay relationship. Understanding soil is something our agricultural scientists and many farmers stopped doing a long time ago. The diversity of soils is not very different from the diversity of plants or animals. If indeed we are to evolve beyond raping the Earth we have to work with the soil and attempt to understand its needs and special character as farmers were doing very competently across Asia 1000 years ago. For instance in the tropics , the fertility of the soil is stored in the tree canopy which would necessitate a long term cropping strategy that would include trees as an inbuilt mechanism for sustaining soil fertility. Furthermore cropping patterns and agricultural systems in the tropics should attempt to imitate the tropical rain forest ecology and NOT blindly mimic the industrial agricultural models of the temperate regions.

The Green revolution was supposed to help the post colonial world emerge from the hunger crisis. Despite an 800% increase in the world chemical fertiliser use since 1950, there has been only a 53% increase in yield accompanied with a sharp drop in nutritive quality of food, (e.g. a 17% drop in the content of wheat protein has been recorded in the USA); we have not emerged from "the food crisis" despite all the chemicals we have used to boost food production and DESPITE all the hundreds of millions spent to "improve" seeds and agricultural technologies. Even now the dependence of green revolution technology on the use of fossil fuels and the global warming caused thereby, is something that we hesitate to accept and address. Without subsidies, conventional agriculture is economically unviable, besides, it has never paid for its environmental cost anyway.

Even the irresponsible genetic meddling with our food and other crops has really not had any significant breakthrough to justify the expense. We have wasted a lot of time and money playing in the lab and commercial interests have now pushed the unpredictable results of 'playing God' onto our fields and into the unsuspecting environment. The long term effects of this interference in the **global** genome will only show up in the future ... in a totally unpredictable way.

Now we are entering the age of "the energy crisis"; one could simplistically say that food crisis + energy crisis = climate crisis for the human species and many other species as well! But it is more complex than that as we all know.

Making organic farming work; a personal experience

Two and a half decades ago, we (the authors of this paper) belonged to a generation that began to see the Green Revolution go wrong and since then our life has been an experiment to attempt living life by the rules of Nature. Many other organic farming families too are out there to bear testimony that the Earth has a great capacity to recover provided that we humans have the will and humility to start listening to her. Principles of seed-saving, multiple cropping, integrated animal and soil husbandry, utilisation of animal and plant-waste, small-scale food-preservation and storage, backed by strong family participation, has been our path of learning. But it has not been simple. Even our own experience has been a struggle with all the advantages of education and background that we had. We have had to learn not only how to farm per se, but also how to survive as a farming family. It has become clear that the sustainability of our operation depends on the diversity of our crops and animals both from the aspect of establishing a fundamentally stable ecology on the farm as well as in terms of optimizing energy inputs to maximise fertility, bio-mass and crop out-put. Wild species diversity has been progressively widening as well providing food for wild flora and fauna, particularly birds. Around 250 different trees, shrubs and weeds have been identified and the list grows.

The cultivable area of our farm, which covers about 30 acres of land, grows, besides hedgerows and wild species of shrubs and trees, about 30 different kinds of crops, to which we value-add on-farm in one way or another. On a regular basis, we employ about six persons besides our selves and sell produce to at least 100 families locally in Mysore and other cities. This we do through year-round cultivation of agricultural and horticultural crops.

Almost all of our inputs are produced on- farm such as biodynamic-compost, Panchagavya and our own formulations of soil conditioners. We depend heavily on green manuring and cropping systems which utilise different yet mutually beneficial types of crops including horticultural and plantation crops. In most cases we do not grow more than two acres of a particular crop design at one time. At all times we attempt to keep the soil covered by practising inter-cropping with short-term crops like pulses, especially in cash crops.

Multi crop-cotton, corn, chilli, bhendi and marigold.

Crop-protection is now a thing of the past; it has been our experience that organic farming in itself is a systemic way to correct pest imbalance: frequently pest problems are easier to control because the diversity of plants diffuses the situation quickly. Along with this we attempt to plan our sowing with respect to moon-phase activity and known pest peak periods which helps to reduce critical situations.

As much as possible, we try to use seeds that are grown by other organic farmers or ourselves because the seeds produced for conventional farming are designed to respond to chemical inputs and have poor keeping quality. Furthermore established organic seeds are live material and are fine tuned micro-climatically so that they outperform chemically bred seeds. For instance, there are old sorghum varieties in Karnataka that continue to grow in drought conditions but also have the capacity to suddenly adjust to heavy rainfall; this variety grows up to the height of 12 feet and produces excellent grain, providing both fodder for animals and food for humans. It has been our experience that traditional seed breeding uses criteria that produce crops with a tolerance for diverse weather conditions and require less water while still producing a phenotype that balances bio-mass and food quality. In India the seed laws and seed industry have done their share of harm to mycorrhiza by insisting on all commercial seed to be coated with extremely toxic fungicides which wreak havoc in biologically active soil.

Traditional sorghum seed grown under organic farming.

Weed-control, inter-cultivation and successive planting using manual, mechanical and powered tools have proven to be an effective combination. Shallow tillage, whether bullock-drawn or using a small tractor, has conserved our soil and re-habilitated its natural self-renewing mechanisms. Practical realities are the best motivation to constantly innovate and develop better techniques.

Every year we find the symbiotic relationship between the soil, crops, animals and humans requires less effort to maintain. This is proof of sustainable design; that no one part deprives another part of its ability to access the basic requirements of healthy living. Surviving as an organic farming family has given us the impetus to help other much smaller farmers in our area to adopt organic practices. This has led to greater self-reliance and a higher quality of life. We have worked hard to develop a strong consumer base and thus, the circle of sustainability widens. Earth-centred economics is starting to work for some of us.

In conclusion...... Evolving a strategy for Planet Earth

The differences between Organic Farming and high-input chemical intensive farming is a useful way to understand more clearly the direction of development taken by the human species in the past 75-100 years as contrasted with the direction we COULD take. The path we were on when science was kept within the boundaries of empiricism based on Natures' model for 1000s of years, in comparison to the path wherein the human race attempts to go beyond this model and attempts to re-create the world to suit its ambitions. The result has been climatically catastrophic because, whether we like it or not, the Earth has finite resources and we cannot exploit them infinitely. As a species we have become wasteful but this was not always so.

Organic farming evolved from the needs of our ancestors who based their agriculture on utilizing what they could harness, with the knowledge they had, to satisfy the requirements of living. Its history is deeply inter-twined with our march towards civilization. In its new Avatar, founded on a holistic view of the Earth and supported with a vast practical experience, may lie some of the solutions we are seeking. There is no doubt that the energy-excessive culture of this Modern age has taught us much about our need to reign in our megalomaniac tendencies. Faced with the opportunity of using the self-contained and renewable energy cycle of organic farming to help regulate and modify the climate of the Earth are we going to have the courage to let go being wasteful and instead and do what is right ? Organic farming offers practical and plausible options to not only slow down the rate of climate change but also adequately feed the world population of humans and their biological allies who all belong equally to the Earth.

References

24 years of practical experience In Organic farming.

Sustainable Farming Systems , by Vivek and Juli Cariappa 2000.

Regenerating the Soil, Claude Bourgignon.

The Gaia Atlas.

The Biodynamic farmers handbook, Norrie Pearce.

12

Indian Ocean and Climate

D.R. Sikka

Introduction

Earth's climate system is composed of several interacting components, like ocean, atmosphere, biosphere, cryosphere and geosphere. The interactions occur on different spatial and temporal scales. Spatial scales vary from global, regional, sub-regional to local, and temporal scales vary from thousands of years, hundreds of years, decades, inter-annual and sub-seasonal, to daily and hourly to smaller time scales. There are several feed back processes opposing or helping each other within each component of the climate system as well as within the interacting loops between different components of the system.

The mobile parts of the climate system are the ocean and the atmosphere. The ocean is the slower and the atmosphere is the faster acting component. The climate system is powered by the incoming solar energy. Most of the solar energy is received at the surface of the earth as atmospheric gases absorb only a fraction of the incoming short-wave solar radiation. The atmosphere is heated from below by conduction and convectional processes and the heating is achieved through long-wave emission from the earth's surface — land and ocean. The ocean is heated from top to bottom and hence the heating process is not very efficient. The temperature of deeper water does not change much in the annual cycle, whereas the annual cycle of temperature in the case of the tropospheric part (first 10-16 km above the earth) of the atmosphere changes more rapidly.

Atmospheric density is low and decreases rapidly with height. Ocean is a denser medium and its density is controlled in a complex way by its salinity and temperature distribution. Ocean transports heat energy to atmosphere as well as to deeper parts of the ocean, and some in the meridional and zonal directions through ocean currents. Atmosphere exchanges heat and momentum at its lower boundary with the ocean, and the wind force drives the upper ocean circulation. Ocean drives the atmosphere by exchanging sensible heat and latent heat of vaporisation at the ocean-atmosphere interface. The latent heat is carried by the water vapour horizontally to great distances (thousands of kms) which later condenses as clouds and releases the latent heat in areas of precipitation. The latent heat provides the motive power for maintaining the atmospheric weather systems. The land and the geosphere also receive solar energy, and transport heat and moisture to the overlying atmosphere.

The land and biosphere are slow parts of the climate system in some respects and quick responding parts in some other aspects. Thus, a highly complicated pattern of coupled motions, thermal contrasts and interactions take place within different components of the climate system which are responsible for maintaining a long-term balance in the system while short-term climatic and weather fluctuations do take place on various spatio-temporal scales. The links between different components are intricate through vital to the operation and understanding of each component as well as the climate itself on global, regional and sub-regional scales. In this paper, we are mainly concerned with the links between the Indian Ocean and the atmosphere in operating the climate over the Indian Ocean.

Section II eleborates upon the general aspects on ocean-atmosphere interactions and climate.

Sections III discusses the broad seasonal climatic regimes of India, focuses on the geographic position of India *vis a vis* land-ocean distribution and emphasises the monsoon climate of India.

Since summer monsoon is the dominant climate regime of India with respect to water and agricultural resources of the country, the role of Indian Ocean, Arabian Sea and the Bay of Bengal is discussed in Section IV emphasising the temporal evolution of the Indian summer monsoon.

As seas around India spawn cyclonic storms which strike its coastal belt, some aspects of the cyclonic storms in the Indian seas are also discussed in Sections III and IV. On the sub-seasonal and interannual scales, the coupled ocean-atmosphere system over the Indian Ocean is linked with the adjoining Pacific Ocean basin and the linkages between the two basins in respect of tropical disturbances on sub-seasonal scale and El Nino - Southern Oscillation (ENSO) - Monsoon connections on inter-annual to decadal scales are touched upon in Section V.

II

Ocean-atmosphere interactions and climate with special emphasis to the Indian Ocean

At the macro-level, the large scale atmospheric wind systems are driven by the thermal contrasts between the poles and the equator, and water vapour acts as a fuel by which process the latent heat is converted as usable heat through clouds and precipitation. The surface winds, in turn, drive the ocean currents which transport almost equal amount of heat as the wind systems from equatorial areas to polar areas. Thus, the ocean returns part of the solar energy to the atmosphere through evaporation and heat exchanges over warm waters. Evaporation rate or the flux of latent heat into the atmosphere affects the atmospheric circulations by altering atmospheric energy source which, in turn, affect the wind and wind stress in the ocean and hence in the motion state of the ocean. Sensible heat is exchanged by the ocean to atmospheric boundary layer where the ocean is warmer than the atmosphere, and from the atmosphere to the ocean where the surface layer of the atmosphere is warmer than the sea surface temperature (SST).

The exchanges between the ocean and the atmosphere are basically turbulent in nature as the two are being heated differentially at variety of spatial scales ranging from tiny eddies to planetary scale gyres supplying, and removing, energy from each other. Besides, the motion in the two are concentrated in certain energetic regimes in terms of strong winds and currents. The Indian Ocean is one of the energetic areas where the giant Asian Monsoon systems both in summer and winter seasons, are located and are known as the Asian summer and winter monsoons. The Indian summer monsoon (June-September) is the dominant part of the larger Asian summer monsoon and provides nearly 80 per cent of the annual rainfall over India. The winter monsoon over India is known as the North East Monsoon, and is responsible for the rainfall over much of the Tamil Nadu, South Karnataka and Kerala states during October to December.

Modes of ocean-atmosphere interactions

Atmosphere is heated from below and vertical scale of convection in the atmosphere extends from surface upto the troposphere (10-16 km), while deep convection occurs over continents as well as over warm tropical oceans. Ocean is heated from top downwards and depth scale of oceanic convection is small, mostly confined to the thermocline (300 m). The oceanic thermocline (temperature and salinity circulation), which affects the deep ocean, is rather slow.

Most of the energy used by the atmosphere is available in the near-equatorial zone of low pressure trough, also known as the Inter-tropical Convergence Zone (ITCZ), as well as in the mid-latitude storm track zone (40°-60° N/S). The conversion of energy is direct in the tropics (30° N - 30° S) such that warm air rises and cold air sinks generating kinetic energy from the available potential energy. This overturning mostly takes place in the meridional plain (south to north in the northern hemisphere), and partly the overturnings are in the zonal (west to east) plane also. The meridional overturning in the large scale is known as the Hadley Circulation and the zonal overturning is known as Walker Circulation. Both these forms of large scale overturnings are important in the context of the climate of India. The convective overturning in the mid-latitude belts, in contrast to the tropical atmosphere, are indirect such that cold air is forced to rise and warm air subsides.

The upper part of the tropical ocean is the major source of energy and moisture to the atmosphere. The Indian Ocean, Western Pacific, the eastern part of the north tropical Pacific and the north tropical Atlantic Ocean are warm and the Surface Sea Temperature (SST) is mostly above 26°C in these oceanic regions. The warm pool of the tropical ocean lies from 20°N-20°S from 60°E to 180°E over the Indo-Pacific belt and, over certain portion of this oceanic belt, the SST is 29°-30° C during most of the year. The warm pool in the north Indian Ocean extends from south-east Arabian Sea to north Arabian Sea and covers the Bay of Bengal which is recognised as an important energetic zone.

A special feature of the tropical ocean is that it can adjust to wind changes rather fast in comparison to the high latitude ocean. The ocean, like the winds, is influenced by hydrological processes such as precipitation and evaporation. Thus, evaporation increases surface salinity and precipitation reduces surface salinity. These processes are very important in the context of the north Indian Ocean such that even though the Arabian Sea is a vast evaporating area in summer, because of lack of precipitation over central and western Arabian Sea during June to September period, the surface salinity in the Arabia sea is kept fairly high. On the contrary, even though there is substantial evaporation over the Bay of Bengal in summer, because of large scale convection over this area and the associated rainfall over the Bay as well as over the adjoining continental belt, there is large flux of fresh water (introduced through rainfall and run off from several rivers) which keeps the salinity of the Bay of Bengal waters quite low as compared to the Arabian Sea.

Processes in ocean-atmosphere interactions

The ocean supplies heat and moisture to the atmosphere and removes part of its momentum. The atmosphere feeds the ocean through SST as the lower boundary condition and the ocean is driven by wind stress (momentum flux) imparted by the atmosphere.

On the diurnal scale, the ocean heating /cooling near the atmospheric surface layers are involved and salinity remains unaffected. On the scale of days (synoptic scale) the precipitation and cloudiness are important in modulating SST such that the SST and surface air temperatures can fluctuate by 2°-3°C during extensive rain episodes and the ocean-atmosphere interactions substantially increase. Typical example of air-sea interactions in the case of seas around Indian land mass is the land-sea breeze system and on the synoptic scale is the weather disturbances such as monsoon depressions/tropical cyclones. On the seasonal scale, the major interaction is the Indian summer monsoon which derives enormous moisture from the south and the north Indian Ocean and pours it as monsoon rains over the length and breadth of India. Sub-seasonal fluctuations in the summer monsoon occur in the form of active-break cycle of the monsoon on 30-40 day scale such that an extended spell (10-20 days) of good monsoon rainfall is followed by an extended spell of deficient monsoon rainfall. On the interannual scale, the interactions occur in the form of excess (strong) monsoon year *vis a vis* deficient (weak) monsoon year over India, and these fluctuations are related to the El Nino - Southern Oscillation (ENSO) cycle. Thus, seasonal scale of the air-sea interactions over the Indian region are dominated by monsoons and variability in its strength on interannual scale is manifested through ENSO - Monsoon connections.

In the tropics, atmosphere may also initiate anomalies in the ocean, but the ocean's influence on the atmosphere is very strong. Again, Indian Ocean presents us with both types of interactions. For example, the forming of the Somali Basin strong ocean current during May to October is a result of the stress of south-west monsoon on the south and north Indian Ocean. Similarly, the large transport of moisture across the north Indian Ocean and its fluctuations result from ocean's influence on the atmosphere. Atmospheric forcing allows warm air to rise in monsoon cloud bands and the resultant rainfall in turn heats the atmosphere and allows the pressure to fall. The heavy rainfall indirectly changes the SST and salinity of the ocean surface which in turn changes atmospheric circulation. Thus, there are loops within loops in the complex ocean-atmosphere interactions over the warm tropical ocean such as the north Indian Ocean sector.

There could be decadal and longer scales of atmosphere - ocean interactions involving the thermohaline circulation which extends into the deep ocean. An example of such large scale interactions is the conveyor belt of the North Atlantic-Indian Ocean-Pacific Ocean which links the basin such that colder waters in the north Atlantic Ocean slowly sink and appear in the Indian Ocean-Pacific Ocean and return to the Atlantic Ocean. Fluctuations in the conveyor belt circulations could change large scale regional/global climate on several decades to several centuries scale. Our information about such decadal and centennial scales of ocean-atmosphere interactions is very meagre and the knowledge gap may be somewhat filled in next two decades or so through observations, theory and modelling researches under different subprogrammes of the World Climate Research Programmes (WCRP).

III

Thermal climatic regimes and role of Indian Ocean

India enjoys a vide spectrum of climate covering snow-bound Himalayan belt, desert of Rajasthan, tropical rain-drenched forests of north-east India and the west coast, tropical forest of Madhya Pradesh, the moist Indo-Gangetic plain, the sub-arid parts of the Peninsular India and the coastal belt of Andhra Pradesh and Tamil Nadu receiving rains both in the south-west monsoon seasons and the north-east monsoon seasons. Though parts of north-west India enjoy temperate type of weather disturbances in the winter, Indian climate can be aptly described as Monsoon climate in which the period June to September is the main rainy season dominated by south-west monsoon and the period October to December falls in the domain of north-east monsoon. The south west monsoon blows from ocean to land and the north-east monsoon blows from land to ocean and adjoining coastal areas. However, because of the dominant influence of summer monsoon on India's life, culture and economy, the country's seasons are divided with reference to the Summer monsoon, usually called the monsoon season. India's climate is, therefore, characterised as:

(i) Winter season (January - February) with mostly clear weather except for the extreme north-west belt where winter-time mid-latitude disturbances bring some precipitation.

(ii) Pre-monsoon season (March-April-May), which is the season of summer heat over most of India with the heat belt slowly migrating from south to north with the march of the sun. Solar heating creates unstable environment and under the influence of tropical and extratropoical migrating weather systems from east to west and from west to east over the Peninsular and central/north-west/north-east India respectively, thunderstorms either isolated or widespread occur sporadically. The organisation of thunderstorm activity is most marked in north-west India (*'Andhis'* or dust storms), over east and north-east India (as Norwesters) under the influence of west to east migratory western disturbances. The Arabian Sea is the moisture source for north-west India and the Bay of Bengal is the moisture source for east/north-east India. Convective systems over south Peninsular India are mostly fed by the moisture source over the Bay of Bengal.

Tropical cyclone is an important severe weather phenomenon in this season and they occur both in the Bay of Bengal and the Arabian Sea. They form in near equatorial area move in a west/north-west direction and strike the Indian coasts. The Bay of Bengal spawns more cyclones than the Arabian sea and these can strike the coasts of Tamil Nadu and Andhra Pradesh during their west/north-west tracks. Some storms recurve and, moving in north/north-east direction, can strike the coasts of Orissa, West Bengal, Bangladesh and Burma. The storms, on hitting the coast, cause enormous damage and particularly those associated with the storm surges cause havoc to human life. The warm waters of the south and central Bay of Bengal and south-east Arabia Sea are the spawning grounds of these destructive storms. The storms forming in the Arabian Sea and moving north/north-west/north-east strike Goa/Konkan/

Gujarat Coasts in India and Pakistan coast in their northward tracks.

(iii) Monsoon season (June-September): This is the main rainy season for India and the role played by the ocean-atmosphere intractions is most vital in this season. Some aspects of this season will be discussed later.

(iv) Post Monsoon season (October-December): By the mid-October, south-west monsoon withdraws from northern, eastern and central parts of India and hence the period October - December is called the post-monsoon season. With the withdrawal of south-west monsoon, a low pressure area forms over the warm waters of the central Bay of Bengal by mid-October which gradually shifts equator-ward with the march of the season. North-east winds blow across Andhra and Tamil Nadu coasts and circulate around the low pressure over the Bay of Bengal bringing rains of the north-east monsoon season for the south Peninsular India. Tropical cyclones again form in the Bay of Bengal and the south-east Arabian Sea in this season and threaten the Indian and adjoining coasts. As in the pre-monsoon season, Arabian Sea and the Bay of Bengal are the moisture sources for these storms and air-sea-interactions dominate the season on 1-5 day and longer 10-20 and 30-40 day scales.

From the above, it can be stated the Indian climatic regimes are vitally influenced by the surrounding oceanic waters and the Himalayan region. The forest belts of eastern and north-east India also interact with the ocean-atmosphere regime. Summer monsoon is basically land-atmosphere-ocean coupled system. The Indian seas (Arabian Sea and the Bay of Bengal) are within the warm pool of the tropical oceans and hence influence the climate on diurnal, synoptic, supersynoptic and interannual scales temporally and spatially in the form of diuranal cycle of convection, synoptic disturbances (depressions, cyclones), extensively organised monsoon rain bands and rains on the scale of India. India is also located to the west of the warm pool of the tropical Pacific Ocean and hence weather disturbances travel west ward from the Pacific Ocean to the Bay of Bengal. They are rejuvenated by the warm waters of the Bay of Bengal.

IV

Summer monsoon of India and its fluctuations

Summer monsoon of India passes through three major phases during its life cycle of four months and the Indian Ocean plays a dominant role in all the phases. These stages are:

(i) Onset and advance of monsoon

(ii) Established monsoon and sub-seasonal fluctuations in the established phase

(iii) Withdrawal phase of monsoon

Onset and advance phase of monsoon

The monsoon onset occurs with the establishment of cross-equatorial flow across the western equatorial Indian Ocean. Air in the form of south-east trades originating in the region of the high pressure area in the sub-tropical South Indian Ocean the equator and becomes south-west monsoon over the western Arabian Sea. The strong airstream in the belt 5°-15°N over the Arabian Sea is known as the low-level Jet and is the main artery through the south Indian Ocean and the north Indian Ocean which transports moisture toward India. Over the Indian landmass, the pressure falls in May and early June and a trough of low pressure develops across the Gangetic valley by mid-June. The low level air flow converges and as such air rises in the belt of convergence. By the end of May, an anticyclonic diverging air stream at the upper troposheric level is established near 20°-25°N. This migrates toward the heated Tibetan Plateau by mid-June to early July. The low pressure region in the lower and middle troposphere lies over warm south-east Arabian Sea and the south Bay of Bengal by mid-May and cyclonic disturbances migrate from east to west bringing pre-monsoon thunderstorms over south Peninsular India. Some of these disturbances intensify into storms. By the end of May to early June, circulation changes occur over India and neighbourhood.

The airflow of the south-east trades rushes northward across equator, converges into the region of intensifying vortex or trough in the mid-troposphere producing heavy rainfall first over the Andaman Islands and subsequently across Lakshdweep Islands and Kerala coast. This is known as the burst of the monsoon over Kerala. The weather disturbance moves northward, taking the heavy rainfall zone along with it toward Karnataka, Goa and Konkan coasts across the west coast of India and Andhra Pradesh, Orissa, West Bengal and north-east India across east coast and further north-east ward. Thus, the initial advance of the monsoon is northward or north-west ward. The advance of monsoon across the Gangetic valley usually occurs from east to west across West Bengal, Bihar, Uttar Pradesh, Haryana, Punjab and Rajasthan. Since the advance is accomplished with the formation of synoptic scale weather systems, each lasting 3 to 5 days, the advance process takes about 30-40 days to be completed for the whole of India, beginning from Kerala and ending in west Rajasthan. The advance is never accompalished in one spell but in 3-4 spells, each spell associated with the formation of a weather system over the Indian seas and migrating over mainland India, and taking the moist monsoon flow with it. Thus, the ocean's role is to provide the moisture to the large scale monsoon and maintain the weather disturbances through release of latent heat over the ascending branch of the weather systems. Synoptic scale disturbances in the onset and advance phases are the low pressure areas, depressions, cyclonic storms and the mid-tropospheric troughs, lows and cyclones. Usually 4 to 5 weather disturbances of varying intensity form in the onset and advance phases which last over 30 to 50 days. There is considerable year to year variability in the accomplishment of this phase. However, the monsoon onset and advance is the most unique feature in the coupled land-atmosphere ocean system during the early phase of summer monsoon over India. Much research work has been done by many workers in understanding the dynamics of this phase of monsoon.

Established phase of monsoon

July and August period is usually considered as the established phase of summer monsoon over India when the monsoon rains have sway over the entire country. The monsoon convergence zone lies over the Gangetic plain running from Rajasthan and Punjab toward the north Bay of Bengal. The western part of the trough is the seat of dry weather with occasional rainspells in association with synoptic scale weather disturbances. The eastern part of the trough lies over the warm waters of the north Bay of Bengal and is usually the seat of convective clouds, formation of weather disturbances on 3 to 5 day scale and heavy rainfall in the south-west sector of the disturbances and relatively less on the northern and eastern sides. These disturbances move into the monsoon trough over the Gangetic valley bringing rain to Bihar, Utter Pradesh, Punjab and Rajasthan. The large scale monsoon trough also fluctuates in intensity and position, and moves north of the mean position with a migratory disturbance and south of the mean position when a fresh disturbance forms over the Bay of Bengal. There are spells of 2-3 weeks at a stretch when the monsoon trough is active south of the mean position with the formation of 2 or 3 overlapping disturbances in the trough zone. This is known as the active phase of monsoon.

When it weakens, the trough shifts toward the foothills of Himalayas and locks itself in that situation for 10 days at a stretch in the western and central parts of the Himalayas.

During such a situation, known as the break monsoon, the rains cease over most parts of India and a secondary trough zone is observed in the near-equatorial region of the Bay of Bengal. The near-equatorial trough once again migrates slowly northward, reaches the north Bay of Bengal in 7-10 days, and the monsoon becomes active again on the central parts of the country. Such spells of excess and deficient rains over the country is known as active-break cycle of the monsoon.

The Indian seas transport tremendous amount of water vapour through latent heat exchange with the atmosphere during active spells and such transport is less during the break monsoon or the weak monsoon spell. The incoming solar radiation over the ocean surface decreases, and latent heat and sensible heat exchanges increase in the good monsoon spells and the reverse occurs during the weak spells. Thus, ocean-atmosphere exchanges are considerably modulated during the active-weak spell of monsoon activity. The ocean loses heat during an active spell with lowering of SST and gains heat during a weak spell. The amplitude of SST oscillation, as observed recently by meteorological ocean buoys in the Bay of Bengal, is found to be 2°C, which is large. Thus, the intra-seasonal oscillation in SST are even higher than the interannual oscillation of SST in the Indian seas. The active-weak cycle of monsoon occurs on 30-40 day scale and, in this process, the convective activity moves from south Bay to north Bay and then to the Himalayas and the cycle begins afresh.

Withdrawal phase of monsoon

As the sun begins to recede equatorward, the land - ocean system responds to the differential heating. The land begins to cool and the monsoon trough weakens and shifts over the warm Bay of Bengal by the beginning of October. The withdrawal of monsoon rains first begins from north-west India and gradually covers central and eastern India. The withdrawal, like the onset, is not accomplished in one shot but in 2-3 spells and, in each spell, the monsoon loses its intensity, moves southward and eastward and the monsoon band stops to migrate further northward from the position it occupied in its previous migration. The Indian seas are still quite warm and the central Bay of Bengal becomes the main area of formation of weather disturbances. The end of the withdrawal phase appears by early October and, by that time, the monsoon trough loses its identity and the seasonal low pressure area positions itself over the central Bay of Bengal with little pressure gradient over mainland India. This heralds the setting up of the post-monsoon season and the beginning of the second cyclonic storm season of the Indian seas as discussed in Section III.

Inter-annual oscillation of the monsoon and ocean-atmosphere interactions

The monsoon rains fluctuate on interannual basis with a standard deviation of about 10 per cent on the India-average basis and higher than this magnitude on sub-seasonal and meteorological subdivisional basis. There are years when the monsoon over India-average basis is more than one standard deviation and they are called Excess Monsoon Years. The reverse of the excess monsoon years in which the All India rainfall is less than one standard deviation are called deficient or drought monsoon years. Thus, on the interannual basis, flood and drought years appear interspersed with the near-normal monsoon rains over the All India scale. Similarly, excess and deficient monsoon years appear on the sub-divisional basis. Since there is considerable homogeneity in the monsoon rainfall distribution on the country scale, such excess or deficient monsoon years cover majority of the sub-divisions simultaneously. Inspite of high correlations between the sub-divisional and All India rainfall, and the high correlation between the neighbouring sub-divisions on the seasonal scale, the excess or drought monsoon years only agree for 50 per cent of the cases on long-term record basis for the neighbouring sub-divisions.

A year may be a good (deficient) monsoon year for the country as a whole, there may still exist deficient (excess) rains over different sub-divisions and districts, or talukas within a sub-division or a district. Thus, the monsoon rainfall has considerable variability on smaller scales and shows even chaotic behaviour as compared to the All India rainfall. This chaotic behaviour is partly due to the fluctuations of the monsoon trough in its position and intensity on smaller scales, and partly due to the tracks of monsoon depressions and lows, over different sub-divisions of the country. Since the monsoon rainfall is organised by large scale weather disturbances with meso-scale convection embedded in them, the chaotic nature of the rainfall distribution on smaller spatial scale on seasonal basis can be explained by the small changes in macro-scale and large scale distribution of individual rain events.

V

Role of Indian and Pacific Ocean in the interannual variability of rainfall and ENSO-monsoon connections

Researchers have investigated different mechanisms for understanding the interannual fluctuations of monsoon rains over India. The variability in interactions between the ocean-land-atmosphere systems is one of the prominent mechanisms for this. These interactions are brought about as the regional scale Asian monsoon system interacts with the Indian Ocean and the adjoining tropical northwest Pacific Ocean. The global scale ineractions are due to the interactions of the Asian monsoon with the equatorial eastern Pacific Ocean known as the ENSO-Monsoon connections.

Interactions of monsoon with the Indian Ocean and the tropical western Pacific

These occur primarily on synoptic (3-7days) and low frequency sub-seasonal scale (10-45 days). The south Indian Ocean is the origin of the monsoon over India and supplies moisture to the air stream. The moisture supports the formation of monsoon disturbances and monsoon trough. Since large amount of energy is supplied by the north Indian Ocean, the Indian seas cool during the evolution of the monsoon—the cooling being large and abrupt at the monsoon onset phase, and sluggish and episodic during the active - break spells of the established phase of monsoon. The neighbouring warm tropical north west Pacific Ocean also spawns tropical disturbances (depressions, storms and typhoons), more in number than the Indian seas. Some of these disturbances migrate west-ward on 10-20 day scale, cross the south-east Asian landmass, enter the Bay of Bengal, re-intensify and finally move over the Indian mainland. Thus, tropical north western Pacific Ocean complements the Indian seas in maintaining the active cycle of the monsoon. Thus, 10-20 days scale weather distubances are generated either over warm waters of the Indian seas or migrate from the adjoining warm western Pacific Ocean along 15°-25°N.

During the active spell of the monsoon, there are episodes when monsoon cloud band extends from western India to east of the Philippines Sea as a single entity extending from 75°E to 140°E with several synoptic scale disturbances embedded within it at the interval of 2,000-3,000 km. The interactions of the monsoon with the north Indian seas and the Bay of Bengal, and the adjoining western Pacific have been the subject of many investigations for over six decades. There is another way by which the equatorial south Indian Ocean interacts with the monsoon as this warm oceanic belt during the monsoon season is also the second seat of organised convection. During active monsoon or excess monsoon, the secondary zone of convection over the equatorial south Indian Ocean is weak. The reverse is the case for the break monsoon or a drought year.

ENSO-monsoon connections

The study of global scale ENSO-Monsoon connections has been pursued for over 70 years since the time of Sir Gilbert

Walker. These connections are now understood in the form of oscillation of the east-west circulation in the equatorial plane in terms of Walker Circulation. Normally, the ascending branch of Walker Circulation lies over Indonesia and the adjoining equatorial Bay of Bengal when the pressure is anomalously low over this region and the SST is anomalously warm. The descending branch of the Walker cell normally lies over the eastern equatorial Pacific where the SST is cold and the pressure is anomalously high over the tropical eastern Pacific. In some other years, the anomalous SST and pressure in the two centres of actions reverse such that the eastern Pacific becomes anomalously warm by 2°-5°C and the pressure over the tropical south Pacific is anomalously low. This warm stage of the eastern equatorial Pacific leads to *El Nino* phenomenon and the differences between pressure over the tropical south Pacific and equatorial Indo-Pacific belt as the negative phase of the Southern Oscillation. The combined phenomena is a coupled ocean-atmosphere phenomena and is known as ENSO. The response of the atmosphere to warm signal of ENSO is observed as the eastward shift in convection from Indonesian-Indian sector to central and eastern equatorial Pacific.

Researchers have linked the warm and cold phases of ENSO (*El Nino* and *La Nina*) to the drought and excess monsoon years respectively over India. Although most of the drought years in Indian monsoon fall during the warm ENSO phase. The non-unique association of monsoon with ENSO is the current subject of active research and it is increasingly believed that monsoon is a complex land-atmosphere-ocean system and is interacting with other components of the climate system such as Eurasean snow cover in winter / spring and continued heating and distribution of soil moisture over the Indian sub-continent.

On the decadal scale, the ENSO-Monsoon connections are also found to be somewhat related — there are more frequent monsoonal droughts (floods) during extended warm ENSO periods (cold periods) on decadal scale. The exact relationship is again very complex and is yet to be understood. There are views about the ENSO leading to monsoon anomalies, and opinions are being expressed that anomalous monsoon may even impact on the evolution of the ENSO cycle itself.

The ENSO-Monsoon Connections are highly tangled as two large scale participatory systems have variability from other factors too. The forthcoming international climate-related programmes like Climate Variability (CLIVAR) and Indian Climate Research Programme (ICRP) would certainly enhance our knowledge about the ENSO- Monsoon connections. This would be possible by studying more intensely the monsoon part, the ENSO part and the coupled ENSO-Monsoon part in the next two decades.

VI

Research programmes on the monsoon

Past experiments

Since the 1960's, several international and bilateral experiments have been undertaken to understand the monsoon and its variability on different spatio-time scales. These experiments are:

(i) International Indian Ocean Experiment (IIOE) (1963-65).
(ii) Monsoon - 73 (1973)
(iii) Monsoon - 77 (1977)
(iv) MONEX - 79 (1979)
(v) INDEX (1980's)
(vi) TOGA (1985-95)
(vii) Monsoon Trough Boundary Layer Experiment (MONTBLEX) (1989-90).
(viii) INDOEX (1997-1999).

These experiments have deployed a wide variety of observing systems—atmospheric, ocean based and space-based research ships and aircraft and satellites have been used to understand the monsoon variability and role of the ocean in that. Our knowledge about the monsoon variability has tremendously increased as a result of studying the data collected during these experiments and numerical modeling, using computers. Many research papers have been published by Indian and international ocean-atmosphere research community.

Indian climate research programme (ICRP)

The Indian Ocean atmosphere research community has proposed a major programme ICRP to understand climate variability over India and its possible prediction. The programme envisages observational and theoretical modeling with intensified vigour. It also envisages carrying out of special process-oriented field programmes to understand land-ocean-atmosphere coupling during the monsoon season. The field programmes would deploy ocean-observing ships, satellites, ocean buoy network in the Indian seas, environmental atmospheric observing system to collect data over the Bay of Bengal, Arabian Sea and the Indian mainland. Bay of Bengal Monsoon Experiment (BOBMEX) will focus on the sub-seasonal scale ocean-atmosphere interactions, controlling the seasonal monsoonal fluctuations. Arabian Sea Monsoon Experiment (ARMEX) will explore the warming of the south-eastern Arabian sea in the pre-monsoon season, cooling in the monsoon season and the role of the Arabian Sea off the west coast of India in heavy rainfall spells over the coast during monsoon. The land component of Indian monsoon is involved in both the above experiments. However, it is planned to carry out a separate experiment on the landmass of India to understand the role of land-ocean-atmosphere interactions in monsoonal fluctuations. A fourth process-oriented study is exclusively focussed an understanding the ocean circulation and thermal variability during the course of a year over the Bay of Bengal and the Arabian Sea.

ICRP has another dimension which emphasises the impact of monsoon variability on societal systems in India—particularly with regard to variability in agricultural production. Researches of the Indian Council of Agricultural Research (ICAR) are expected to play a dominant role in linking the agricultural production system with monsoon and climate variability over India.

The future of research under ICRP in understanding and predicting ocean-atmosphere connections for climate variability over India and modulating agricultural production in the country is quite promising.

13

Climate Change and Agriculture

Surinder Sud

A formidable challenge to Indian agriculture comes in the form of global warming and the consequential climate change. The Indian climate has already undergone a perceptible change and *its* impact on agriculture has begun to crystallise. This is evident from the fact that the annual mean temperature in the country has remained consistently above normal since 1993 with 1997 being the only exception till 2007. The wheat crop has begun to manifest the adverse effect of this-its output remained depressed for several years till 2006 due to the early onset of summer. Other crops, as well as other sectors of agriculture and *its* allied fields, cannot remain unaffected for long.

The *Annual Climate Summary, 2007,* released by the India Meteorological Department (IMD) in April 2008, contained several startling facts, providing clear evidence of the impact of global warming on the Indian climate: 2007 was the fourth warmest year on record since 1901; the annual mean temperature in the country as a whole was recorded at 0.55 degrees Celsius higher than normal (long period average temperature). Notably, all the three years warmer than 2007 have been after 2000. The warmest year in India so far was 2002. It was followed by 2006, 2003 and 2007 in that order. Of the 10 warmest years since 1901, eight have been after 1998. These were 2002, 2006, 2003, 2007, 1998, 2004, 1999, and 2001, in the order of warmness. The two years earlier than 2000 to figure in the top 10 warmest years were 1941 and 1958. The IMD report categorically stated that the year 2007 witnessed extreme weather events across the country, both in terms of temperature and precipitation. During the year, the country experienced severe cold spells, heat waves, floods as well as drought in different parts, which resulted in the death of over 2,000 people. The oceans around India also experienced some of the most severe tropical storms and super cyclones.

Though the total rainfall in the country during the 2007 monsoon season (June-September) was 106 per cent of the normal, the post-monsoon rainfall (October-December) was rather poor, just 68 per cent of the normal. Besides, even during the monsoon season, moderate drought conditions (rainfall deficiency ranging from 26 per cent to 50 per cent) prevailed in the key agricultural belt spanning from the north-west to central India. The affected areas included west Uttar Pradesh, Haryana, Delhi, Chandigarh, Punjab, Himachal Pradesh and east Madhya Pradesh.

Some of the studies conducted earlier by the IMD and the Indian Institute of Tropical Meteorology, Pune, also point to significant changes in temperature and monsoon rainfall in different parts of the country. They reveal a discernible increasing trend of rainfall along the west coast, north Andhra

Pradesh and north-west India and a decreasing trend in east Madhya Pradesh and adjoining areas, north-east India and parts of Gujarat and Kerala (minus 6 to minus 8 per cent over the last 100 years).

The surface air temperature in the last century (1901-2000) indicated a significant warming up to the extent of around 0.4 degrees Celsius. The trends of temperature in different regions indicated a significant warming up along the west coast, central India, the interior peninsula and north-east India. However, some cooling has been observed in the north-west and some parts of southern India. Besides, there is evidence that the glaciers in the Himalayas are receding at a rapid pace.

According to the future projections made by these India-specific studies, monsoon rainfall is expected to increase by 15-31 per cent by the end of the 21st century. The mean annual temperature is anticipated to rise by 3-6 degrees Celsius. The maximum increase in temperature will be in northern India. The warming up is projected to be felt more in winters and the post-monsoon seasons than in the rest of the year.

Globally, 11 of the 12 years previous to 2007 ranked among the 12 warmest years on record since 1850. The mean earth temperature has changed by 0.74 degrees Celsius between 1906 and 2005. Most of the observed increase in global average temperatures since the mid-20th century is attributed largely to the enhanced anthropogenic (human activity related) green house gas emissions.

This apart, in the past 50 years, cold days, cold nights and frost formation have become less frequent while hot days, hot nights and heat waves have become more frequent globally. The rate of occurrence of heavy rainfall has spurted over most land areas during this period. As a result, global sea level rose at an average rate of 1.8 millimetres per year between 1961 and

2003. Worse still, the rate of rise of sea level has surged substantially to 3.1 millimetres a year after 1993.

The future projections in this context are even more worrisome. The anticipated temperature increase by the end of the present century is reckoned to be in the range of 2-4.5 degrees Celsius. Even the most conservative estimate puts it at about 3 degrees Celsius. But the rise, in any case, is unlikely to be less than 1.5 per cent, though the chances of it increasing beyond 4.5 per cent are not being ruled out. For the next two decades, warming of the order of around 0.2 degrees Celsius per decade is considered quite likely.

Also, future tropical cyclones are forecast to become more intense, with higher peak wind speed and more heavy precipitation (water downpour). The frequency of hot extremes, heat waves and heavy rainfall events is foreseen to increase. Perceptible rise in the amount of precipitation is projected for high latitudes while a decrease in rainfall seems in store 'for most sub-tropical land regions. The sea level is projected to rise by 0.18-0.59 metres. Moreover, the pH value (measure of salinity and alkalinity) of the global ocean surface is anticipated to drop by between 0.14 and 0.35 units in the 21st century. This essentially means a further rise in the sea water salinity.

The likely economic fall-out of the global warming and its impact on the world food security and commodity prices has been brought out, even if briefly, in the biannual overview of the world food situation presented by the Washington-based think-tank, International Food Policy Research Institute (IFPRI) in December 2007: 'Climate-change risks will have adverse impacts on food production, compounding the challenge of meeting global food demand. Consequently, food import dependency is projected to rise in many regions of the developing world. With the increased risk of droughts and floods due to rising temperatures, crop yield losses are imminent. In more than 40 developing countries -mainly in Sub-Saharan Africa-cereal yields are expected to decline, with mean losses of about 15 per cent by 2080 .... Other estimates suggest that although the aggregate impact (of climate change) on cereal. production between 1990 and 2080 might be small-a decrease in production of less than 1 per cent-large reductions of up to 22 per cent are likely in south Asia.' This bears out the apprehensions that the South Asian region may be worst hit from global warming and climate change. On the other hand, developed countries and Latin America are expected to experience absolute gains.

The effect of climate change on the production of cereals differs in different crops. Projections show that land suitable for wheat production may almost disappear in Africa. Nonetheless, global land use is estimated to increase minimally, by less than 1 per cent, due to climate change. In many parts of the developing world, especially in Africa, an additional 8 per, cent land may turn arid by 2080.

Significantly, world agricultural GDP is visualised by the IFPRl to decrease by 16 per cent by 2020 due to global warming. Again, the impact on developing countries will be much more severe than on developed countries. As such, the output in developing countries is projected to decline by 20 per cent, while in industrial countries it is projected to decline by only 6 per cent. Carbon fertilisation (benefit of higher carbon dioxide content) could limit the severity of climate change effects in these countries to only 3 per cent.

Equally notably, the IFPRl has cautioned, rather categorically, that the technological change is not expected to be able to alleviate output losses and increase yields to a rate that would keep up' with growing food demand. Agricultural prices will, thus, also be affected by climate variability and change. 'Temperature increases of more than 3 degrees Celsius may cause prices to increase by up to 40 per cent,' IFPRI maintains.

Impact

The changed climate is bound to alter the agro-climatic conditions for crop farming as well as livestock rearing in most parts of the country. The apprehended disappearance of the river-feeding glaciers may dry up flows in the country's major rivers, including the mighty Ganga, and adversely impact agriculture to the detriment of the poor small and marginal farmers who may lack the wherewithal to adjust to the changed climate.

Theoretically, these changes in climate can affect, to a considerable extent, crops, soil, livestock, fisheries and pests. The effects of such changes can be manifested in several ways, such as reduction in crop duration (resulting in early or premature grain ripening); increase in respiration rates of plants and evapotranspiration (leading to higher moisture loss, necessitating more intensive irrigations); disturbance in the equilibrium between crops and pests (pushing up losses due to pests and diseases); and hastening of nutrient mineralisation in the soil (causing higher losses of applied fertilisers)

Also, the changes in temperature and sea level can affect fisheries directly as well as indirectly through the possible modifications in fish species and available feed. Where other livestock are concerned, higher temperature and changed fodder and water scenario can influence the production of meat, milk and poultry products.

There is, however, a silver lining, though thin and intangible. It is thought that the anticipated increase in the atmospheric carbon dioxide content will improve photosynthesis (conversion of solar energy into economic produce) in some crops and boost their productivity and production. This may, to an extent, benefit crops like wheat, rice, legumes (pulses) and oilseeds. However, other crops like maize, *jowar* (sorghum), *bajra* (pearl millet) and sugarcane may not benefit from the increased presence of carbon dioxide. In any case, the overall negative stimulus may overwhelm the positive forces, necessitating major changes in agronomic practices.

As mentioned earlier, the wheat crop has already become a victim, causing fresh worries about the country's food security. In the future, too, temperature variations during the wheat growth period can significantly influence its yields. Recent studies conducted by IARI have indicated the possibility of around 4-5 million tonnes loss in wheat production with every rise of 1 degree Celsius in temperature throughout the growing period of the crop, despite some anticipated positive effect on yields due to the likely increase in atmospheric carbon dioxide content. Indeed, the actual loss of output may even be higher than this because these studies are based on the assumption that irrigation availability would continue at the present level, which may actually not be the case.

However, these studies, notably, have also shown that the projected losses in wheat output could be contained through suitable changes in agronomic practices and the use of better technology. Losses can be brought down from the anticipated 4-5 million tonnes to merely 1-2 million tonnes if most of the farmers manage to change the crop sowing time and switch over to the varieties better adapted to the new climatic conditions.

The United Nations' Intergovernmental Panel on Climate Change (IPCC), which has been awarded the Nobel peace prize, has also made an assessment of the likely impact of the global warming on agriculture. It has endorsed the notion that the brunt of the adverse fall-out of climate change would have to be borne in India by the agriculture sector. This is regardless of the fact that the contribution of this sector to greenhouse gas emissions is relatively meager, estimated by the Environment Ministry at 28 per cent only. What is really disquieting is that the IPCC's reckoning has indicated the probability of 10-40 per cent loss in crop output in India due to anticipated rise in temperature by 2080-2110.

Apart from the direct impact, there is the indirect impact. For instance, land use patterns can undergo a significant change in response to factors such as snow melting and availability of irrigation water; frequency and intensity of inter- and intra-seasonal droughts and floods; changes in soil organic matter content; soil erosion; decline in arable land due to submergence of coastal areas under sea water; and availability of energy. All these variations can have tremendous bearing on agricultural production and, hence, food security. In fact, several significant determinants of food availability, such as government policies, capital availability, prices, infrastructure, land reforms and trade policies, are also expected to alter in response to environmental changes.

This problem is compounded by the fact that the population pressure on agriculture is steadily increasing. The Indian population is estimated to swell to around 1.6 billion by 2050, out-numbering that of China. The higher demand for food, fuel, fibre and fodder would have to be met under relatively more arduous circumstances.

Combating Global Warming

Fortunately, Indian agricultural scientists are unlikely to be caught off-guard as they have already begun preparing themselves for the emerging challenges due to climate change. A national conference on climate change and Indian agriculture was organised by the Indian Council of Agricultural Research (ICAR) in Delhi in October 2007, where the focus was on prioritising research and development strategies and preparing an action plan to combat climate change. This meet made several recommendations aimed broadly at adapting agricultural practices to possible climate changes, curbing the forces leading to global warming, and building up capacity to deal with the emerging contingencies. This would broadly involve taking cost-effective measures to reduce the generation of greenhouse gases and increase carbon sequestration; identification and incorporation into the crop plants of genes that can provide them the capacity to withstand stresses posed by floods, droughts and changing climatic patterns; improve the photosynthetic capacity of crops to make the best use of available solar and other forms of energy; and discovering new agronomic practices and cropping sequences that do well in the emerging weather cycles.

Significantly, the ICAR has prepared an elaborate action plan for facing climate change which, prima facie, seems quite reassuring, especially because some of the activities proposed to be taken up under this plan have already been initiated. An inter-institutional project on climate change, involving some 15 research institutes and state agricultural universities, has been launched for critical research on climate change related issues concerning crops, livestock and fisheries. More organisations are proposed to be roped in to expand the ambit of this research programme. Besides, climate change has been identified as a thrust area for research under the new National Agriculture Innovation Project (NAIP) which has replaced the World Bank--supported National Agriculture Technology Project (NATP) on its completion in 2005.

The ICAR has conceived several other initiatives as well to blunt the impact of climate change on Indian agriculture. These include enhancing the capacity for agriculture-oriented seasonal weather forecasts to facilitate preparation of contingency plans for likely temperature and rainfall regimes.

This will help make timely arrangements for seeds of suitable crop varieties and other required farm inputs to implement 'these plans. On the research front, efforts are proposed to be initiated to convert C3 crop plants (relatively meager utilisers of energy and nutrition) like those of wheat and rice to C4 (better utiliser) types, like those of sugarcane and maize, for boosting the inherent production potential of important crops.

The research areas for mitigating the impact of climate change include enhancing the efficiency of the use of nitrogenous fertilisers and water and measures for reducing the emission of greenhouse gases in the agriculture and livestock sectors. From this viewpoint, research on coating of urea with *neem* or the use of *neem* cake for curtailing the release of nitrous oxide (a highly environment damaging greenhouse gas) can be rewarding. This aside, ICAR has suggested to the government fertiliser pricing policies with in-built incentives for the fertiliser producers to churn out slow nutrient releasing products to reduce losses in the form of gases as well as through leaching. Well-conceived modifications in water management policies for paddy cultivation can have a bearing on the release of methane from these fields, besides saving scarce water.

Indeed, some noteworthy successes have already been achieved in a few of these areas, especially in neem-coated urea and slow-release fertilisers. Work is also on to develop better water management schedules for paddy fields. But all this is in the initial stages and requires more research efforts and validation in the fields. Of course, the technologies that have already been tried out successfully in the fields need to be promoted for wider adoption.

In the livestock sector too programmes have been initiated for finding ways of reducing greenhouse gases generated by animals, especially the ruminants like buffaloes which abound in India. A notable recent success in this field is the discovery of a new fungus, scientifically named *Cyllamyces icaris sp,* which has the ability to degrade dietary fibre, thus reducing the generation of methane in the animal's stomach. By doing so, it also helps improve the utilisation of fibrous crop residues, like straw and stubbles, which are used widely in the country as animal feed even while mitigating greenhouse gases discharges from the livestock sector.

This apart, another significant research and development initiative focuses on agroforestry, which can· go a long way in reversing the climate change forces. Agroforestry, involving a combination of trees and crops either together or in sequence, is now being increasingly recognised globally as having the potential to serve as a carbon sink to reduce the load of harmful gases in the environment. Under this initiative, the ICAR entered into an agreement with the International Centre for Research in Agroforestry in 2007 for collaborative research on farm forestry, aimed specifically at dealing with the climate change issues.

Significantly, the new collaborative agroforestry project will endeavour to develop environment-friendly technologies which the small and resource-poor farmers, in particular, can easily adopt. It will evolve novel agroforestry systems which may require introduction of new tree species, besides promotion of the known ones, having good potential for sucking in carbon dioxide. These systems, moreover, will have the capability to adapt to emerging conditions and mitigate the climate change process. The new plant species will, of course, be introduced after due diligence about their compatibility with agroecology and prevailing cultivation practices. This will be ensured by undertaking the complete life cycle analysis· of the new agroforestry systems before introducing them.

Productive agroforestry systems are deemed particularly useful for the north-east, especially for the areas where the environmentally disastrous *'jhum'* (shifting cultivation) is still in vogue. This can enable the nomadic tribes, who cut the forests to grow crops for a while before moving on to new areas, leaving almost bare patches, lead a more settled life by adopting agroforestry which can provide them sustainable subsistence.

However, the outcome of this venture will depend largely on the level of adoption of the research-based strategies by the rural communities. The chances of success will, predictably, increase if these ventures can somehow be linked to carbon trading to generate some additional income to be passed on to the small stakeholders as an incentive for continuing with the new systems.

Obviously, if the action plan and other strategies conceived by the country's National Agriculture Research System (NARS), together with the other ideas generated by different forums, succeed in meeting their objectives, Indian agriculture may not only manage to remain largely unscathed by climate change but may even convert this into an opportunity to gain from. But much will depend on the official policies and the funding support from the public as well as private sources for generating new technologies and promoting them for actual application.

14

The Other Climate Changers Why Black Carbon and Ozone Also Matter

Jessica Seddon Wallack
and
Veerabhadran Ramanathan

At last, world leaders have recognized that climate change is a threat. And to slow or reverse it, they are launching initiatives to reduce greenhouse gases, especially carbon dioxide, the gas responsible for about half of global warming to date. Significantly reducing emissions of carbon dioxide is essential, as they will likely become an even greater cause of global warming by the end of this century. But it is a daunting task: carbon dioxide remains in the atmosphere for centuries, and it is difficult to get governments to agree on reducing emissions because whereas the benefits of doing so are shared globally, the costs are borne by individual countries. As a result, no government is moving fast enough to offset the impact of past and present emissions. Even if current emissions were cut in half by 2050 - one of the targets discussed at the 2008 UN Climate Change Conference-by then, humans' total contribution to the level of carbon dioxide in the atmosphere would still have increased by a third since the beginning of this century.

Meanwhile, little attention has been given to a low-risk, cost-effective, and high-reward option: reducing emissions of light-absorbing carbon particles (known as "black carbon") and of the gases that form ozone. Together, these pollutants' warming effect is around 40-70 percent of that of carbon dioxide. Limiting their presence in the atmosphere is an easier, cheaper, and more politically feasible proposition than the most popular proposals for slowing climate change-and it would have a more immediate effect.

Time is running out. Humans have already warmed the planet by more than 0.5 degrees Celsius since the nineteenth century and produced enough greenhouse gases to make it a total of 2.4 degrees Celsius warmer by the end of this century. If the levels of carbon dioxide and nitrous oxide in the atmosphere continue to increase at current rates and if the climate proves more sensitive to greenhouse gases than predicted, the earth's temperature could rise by as much as five degrees before the century ends.

A temperature change of two to five degrees would have profound environmental and geopolitical effects. It would almost certainly melt all the Arctic summer sea ice. As a result, the Arctic Ocean would absorb more sunlight, which, in turn, would further amplify the warming. Such a rise could eliminate the

Himalayan and Tibetan glaciers, which feed the major water systems of some of the poorest regions of the world. It would also accelerate the melting of the Greenland and Antarctic ice sheets, raising the sea level worldwide and provoking large-scale emigration from low-lying coastal regions. Cycles of droughts and floods triggered by global warming would spell disaster for agriculture-dependent economies.

Some of global warming's environmental effects would be irreversible; some of its societal impacts, unmanageable. Given these consequences, policymakers world-wide seeking to slow climate change must weigh options beyond just reducing carbon dioxide, especially those that would produce rapid results. Cutting black carbon and ozone is one such strategy.

Powerful Pollutants

The warming effect of carbon dioxide has been known since at least the 1900s, and that of ozone since the 1970s, but the importance of black carbon was discovered only recently. During the past decade, scientists have used sophisticated instruments on drones, aircraft, ships, and satellites to track black carbon and ozone from their sources to remote locations thousands of miles away and measure and model how much atmospheric heating they cause.

Black carbon, a widespread form of particulate air pollution, is what makes sooty smoke look blackish or brownish. It is a byproduct of incomplete, inefficient combustion-a sign of energy waste as much as energy use. Vehicles and ships fueled by diesel and cars with poorly maintained engines release it. So do forest fires and households and factories that use wood, dung, crop waste, or coal for cooking, heating, or other energy needs.

Black carbon alters the environment in two ways. In the sky, the suspended particles absorb sunlight, warming up the atmosphere and in turn the earth itself. On the earth's surface, deposits of black carbon on snowpacks and ice absorb sunlight, thereby heating the earth and melting glaciers. The Arctic sea ice and the Himalayan and Tibetan glaciers, for example, are melting as much as a result of black carbon as they are as a result of the global warming caused by carbon dioxide. The warming effect of black carbon is equal to about 20-50 percent of the effect of carbon dioxide, making it the second- or third-largest contributor to global warming. No one knows exactly how much warming it causes, but even the most conservative estimates indicate a nontrivial impact. And its large contribution to the melting of glaciers and sea ice, one of the most alarming near-term manifestations of climate change, is well documented.

The ozone in the lower level of the atmosphere is another major contributor to global warming that deserves attention. (This is different from the ozone in the stratosphere, which shields life on earth from the sun's ultraviolet rays.) A potent greenhouse gas, its warming effect is equal to about 20 percent of that of carbon dioxide. Unlike black carbon, which exists as particles, ozone is a gas. Ozone in the atmosphere is not emitted directly but formed from other gases, "ozone precursors," such as carbon monoxide (from the burning of fossil fuels or biomass), nitrogen oxides (from lightning, soil, and the burning of fossil fuels), methane (from agriculture, cattle, gas leaks, and the burning of wood), and other hydrocarbons (from the burning of organic materials and fossil fuels, among other sources).

Most important, black carbon and ozone stay in the atmosphere for a much shorter time than does carbon dioxide. Carbon dioxide remains in the atmosphere for centuries - maybe even millennia- before it is absorbed by oceans, plants, and algae. Even if all carbon dioxide emissions were miraculously halted today, it would take several centuries for the amount of carbon dioxide in the atmosphere to approach its pre-industrial-era level. In contrast, black carbon stays in the atmosphere for only days to weeks before it is washed away by rain, and ozone (as well as some of its precursors) only stays for weeks to months before being broken down. Nonetheless, because both are widespread and continuously emitted, their atmospheric concentrations build up and cause serious damage to the environment.

Although reducing the emissions of other greenhouse gases, such as methane and halocarbons, could also produce immediate results, black carbon and ozone are the shortest-lived climate-altering pollutants, and they are relatively under-recognized in efforts to stem climate change. Reducing the emissions of these pollutants on earth would quickly lower their concentrations in the atmosphere and, in turn, reduce their impact on global warming.

An Easier Extra Step

Another promising feature of black carbon and ozone precursor emissions is that they can be significantly limited at relatively low cost with technologies that already exist. Although the sources of black carbon and ozone precursors vary worldwide, most emissions can be reduced without necessarily limiting the underlying activity that generated them. This is because, unlike carbon dioxide, black carbon and ozone precursors are not essential byproducts of energy use.

The use of fossil fuels, particularly diesel, is responsible for about 35 percent of black carbon emissions worldwide. Technologies that filter out black carbon have already been invented: diesel particulate filters on cars and trucks, for example, can reduce black carbon emissions by 90 percent or more with a negligible reduction in fuel economy. A recent study by the Clean Air Task Force, a U.S. non-profit environmental research organization, estimated that retrofitting one million semi-trailer trucks with these filters would yield the same benefits for the climate over 20 years as permanently removing over 165,000 trucks or 5.7 million cars from the road.

The remaining 65 percent of black carbon emissions are associated with the burning of biomass-through naturally occurring forest fires, man-made fires for clearing cropland, and the use of organic fuels for cooking, heating, and small-scale industry. Cleaner options for the man-made activities exist. The greenest options for households are stoves powered by the sun or by gas from organic waste, but updated designs for biomass-fueled stoves can also substantially cut the amount of black carbon and other pollutants emitted. Crop waste, dung, wood, coal, and charcoal are the cheapest, but also the least efficient and dirtiest, fuels, and so households tend to shift away from them as soon as other options become reliably available. Thus, the challenge in lowering black carbon emissions is not convincing people to sacrifice their lifestyles, as it is with convincing people to reduce their carbon dioxide emissions. The challenge is to make other options available.

Man-made ozone precursors are mostly emitted through industrial processes and fossil-fuel use, particularly in the transportation sector. These emissions can be reduced by making the combustion process more efficient (for example, through the use of fuel additives) or by removing these gases after combustion (for example, through the use of catalytic converters). Technologies that both minimize the formation of ozone precursors and filter or break down emissions are already widely used and are reducing ozone precursors in the developed world. The stricter enforcement of laws that forbid adulterating gasoline and diesel with cheaper, but dirtier, substitutes would also help.

Fully applying existing emissions-control technologies could cut black carbon emissions by about 50 percent. And that would be enough to offset the warming effects of one to two decades' worth of carbon dioxide emissions. Reducing the human-caused ozone in the lower atmosphere by about 50 percent, which could be possible through existing technologies, would offset about another decade's worth. Within weeks, the heating effect of black carbon would lessen; within months, so, too, would the greenhouse effect of ozone. Within ten years, the earth's overall warming trend would slow down, as would the retreat of sea ice and glaciers. The scientific argument for reducing emissions of black carbon and ozone precursors is clear.

A Political Possibility

Reducing emissions of black carbon and ozone precursors is also a politically promising project. It would yield significant benefits apart from slowing climate change, giving governments economic and developmental incentives to reduce them. Reducing ozone precursors, for its part, would have recognizable agricultural benefits. Ozone lowers crop yields by damaging plant cells and interfering with the production of chlorophyll, the pigment that enables plants to derive energy from sunlight. One recent study estimated that the associated economic loss (at 2000 world prices) ranged from $ 14 billion to $ 26 billion, three to five times as large as that attributed to global warming. For policymakers concerned about agricultural productivity and food security, these effects should resonate deeply.

In countries where a large portion of the population still depends on biomass fuels, reducing black carbon emissions from households would improve public health and economic productivity. Nearly 50 percent of the world's population, and up to 95 percent of the rural population in poor countries, relies on solid fuels, including biomass fuels and coal. The resulting indoor air pollution is linked to about a third of the fatal acute respiratory infections among children under five, or about seven percent of child deaths worldwide. Respiratory illnesses associated with the emissions from solid fuels are the fourth most important cause of excess mortality in developing countries (after malnutrition, unsafe sex, and waterborne diseases).

These health problems perpetuate poverty. Exposure to pollutants early in life harms children's lung development, and children who suffer from respiratory illnesses are less likely to attend school. Air pollution leaves the poor, who often earn a living from manual labor, especially worse off. Collectively, workers in India lose an estimated 1.6 to 2.0 billion days of work every year to the effects of indoor air pollution. Reducing black carbon emissions from households would thus promote economic growth and, particularly for rural women and children, improve public health.

Furthermore, both black carbon and ozone precursor emissions tend to have localized consequences, and governments are more likely to agree to emissions-reduction strategies that can deliver local benefits. With carbon dioxide and other long-lasting, far-spreading greenhouse gases, emissions anywhere contribute to global warming everywhere. But the effects of black carbon and ozone are more confined. When it first enters the atmosphere, black carbon spreads locally and then, within a week, dissipates more regionally before disappearing from the atmosphere entirely in the form of precipitation. Ozone precursors, too, are more regionally confined than carbon dioxide, although background levels of ozone are increasing around the globe.

Because the effects of black carbon and ozone are mostly regional, the benefits from reducing them would accrue in large part to the areas where reductions were achieved. The melting of the Himalayan and Tibetan glaciers is almost reason enough for countries in South and East Asia to take rapid action to eliminate black carbon emissions. So is the retreat of the Arctic sea ice for countries bordering the Arctic Ocean. Regional groupings are also more likely than larger collections of countries to have dense networks of the economic, cultural, and diplomatic ties that sustain difficult negotiations. Moreover, both black carbon and ozone can be contained through geographically targeted strategies because many of the sources of black carbon and ozone are largely fixed. And so even if one country in a region seeks to regulate emissions, that country's polluting activities are unlikely to move to another country with less stringent policies-a common concern with agreements to reduce carbon dioxide emissions.

Cleaning Up

So what can be done to curb black carbon and ozone precursor emissions? A logical first step is for governments, international development agencies, and philanthropists to increase financial support for reduction efforts. Although some money for this is currently available, neither pollutant has emerged as a mainstream target for public or private funding. Simply recognizing black carbon and ozone as environmental problems on par with carbon dioxide would make policymakers more inclined to spend development funds and the "green" portions of stimulus packages on initiatives to tackle them. Developed countries could put their contributions toward customizing emissions-reduction technologies for the developing world and promoting their deployment-an important gesture of goodwill that would kick-start change.

Regardless of the source of the funding, aid should support the deployment of clean-energy options for households and small industries in the developing world and of emissions-reduction technologies for transportation around the world. This could mean distributing solar lanterns and stoves that use local fuel sources more efficiently or 'paying for small enterprises to shift to cleaner technologies. The specific fixes for small-scale industry will vary by economic activity-making brick kilns cleaner is different from making tea and spice driers more efficient-but the number of possible customers for the new technologies offers some economies of scale. When it comes to transportation, policy options include subsidizing engine and filter upgrades, shifting to cleaner fuels, and removing the incentives, created by government subsidies that favor some fuels over others, for adulterating fuel and for using diesel.

Deploying technologies to reduce emissions from so many culturally embedded activities, from cooking to driving, will not be easy. Enforcing emissions controls on many small, mobile polluters is harder than regulating larger sources, such as power plants. And in customizing technologies, close attention will need to be paid to the varied needs of households and industry. But creating and enforcing regulations and subsidizing and disseminating energy-efficient technologies are challenges that have been met before. The "green revolution"-the remarkable growth in agricultural productivity that occurred in the second half of the twentieth century-introduced radical changes to small-scale farming. Other development initiatives have influenced fertility, gender equality, schooling, and other household decisions more sensitive than those about cooking and driving.

Moreover, the infrastructure for inter-national financial and technological transfers already exists in the form of the World Bank, regional development banks, and UN programs that have supported development around the world for decades. The Global Environment Facility, a development and environmental fund that started as a World Bank program and is now the world's largest funder of environmental projects, is well suited to finance cleaner technologies.

Governments and international agencies should also finance technology that tracks air quality, which is generally under monitored. In the major cities of most developing countries, the number of sensors has not kept up with the growth in population or economic activity. In rural areas, air pollution is not tracked at all. Improving the monitoring of air quality and disseminating the data would inform policymakers and environmental activists. And tracking individuals' emissions -through indoor air-pollution monitors or devices attached to cars' tailpipes-could help motivate people to curb their emissions. Experimental initiatives to measure individuals' carbon footprints and energy use have been shown to change people's behavior in some settings.

Aid alone will not be enough, however, International organizations must also help governments identify and act on opportunities that mitigate climate change and promote development. International development institutions, such as the UN Environment Program and the multi- lateral and regional development banks, could sponsor research, set up inter-ministerial working groups, and establish standards for monitoring and reporting public expenditures. These initiatives would make it easier to identify possible areas of coordination among public health, agricultural, environmental, and anti-poverty programs. In most countries, domestic institutions are not designed to encourage cooperation among different authorities. Pitching the reduction of black carbon and ozone precursor emissions as public health and agricultural policies could help such efforts compete for scarce funds; enabling the clearer calculation of the environmental benefits of development policies would make policymaking more informed. Much in the same way that international development organizations currently support good governance to improve infrastructure and services, they should also promote better environmental governance.

Responding Regionally

The current piecemeal approach to climate science-in particular, the tendency to treat air pollution and climate change as separate issues-has at times led to bad policy. The decision of many countries to promote diesel as a means to encourage fuel efficiency, for example, may have had the inadvertent effect of increasing black carbon emissions. And air-pollution laws designed to reduce the use of sulfate aerosols, which cause acid rain, have ironically led to more warming because sulfates also have a cooling effect. Had policymakers instead integrated efforts to reduce air pollution with those to slow global warming, they could have ensured that the reduction of sulfates was accompanied by an equivalent reduction in green-house gases.

A single global framework would be the ideal way to integrate various strategies for mitigating climate change. Bilateral or multilateral agreements are more feasible for getting started on reducing black carbon and ozone precursor emissions. These can strengthen governments' incentives to act by discouraging free-riding and by motivating governments to take into account the larger-scale impacts of their own emissions. Because the sources of black carbon and ozone vary from region to region, agreements to reduce them need to be tailored to suit regional conditions. In the Northern Hemisphere, for example, ozone precursors mostly come from industrial processes and transportation, whereas in the Southern Hemisphere, especially tropical regions, they mostly come from natural emissions (soils, plants, and forest fires). The sources of black carbon vary by region, too: in Europe and North America, transportation and industrial activity playa larger role than the burning of biomass, whereas the reverse is true in developing regions.

The impact of emissions on the climate is scientifically complex, and it depends on a number of factors that have not yet been adequately taken into account when devising climate models. The challenge, then, is to quickly create agreements that consider the complex links between human activities, emissions, and climate change and that can adjust over time as the scientific understanding of the problem evolves. Regional air-pollution agreements are easier to update than global agreements with many signatories. The UN Convention on Long-Range Transboundary Air Pollution (most of whose signatories are European or Central Asian states) and its subsequent pollutant-specific protocols provide a ready model for regional agreements on short-lived climate-changing pollutants. The specific provisions of these agreements are based on the costs of reductions, scientists' knowledge of the sources and distribution of air pollution, and the ability to measure reductions-considerations that should also inform the regulation of black carbon and ozone precursor emissions. Moreover, these agreements commit countries to particular actions, not just specific outcomes. This is wise, given that emissions are difficult to monitor and quantify precisely.

Black carbon and ozone can also be built into existing bilateral discussions. The High-Level India-EU Dialogue, a working group of scientists and policy-makers from Europe and India, is one such existing forum. In February 2009, it was already urging governments from Europe and India to work together to recognize and reduce the threat from black carbon. Participants proposed an interdisciplinary research project that would determine the effects of biomass- based cooking and heating on health and the climate and assess the obstacles to a large-scale deployment of cleaner stoves. Black carbon and ozone are also natural candidates for U.S.-Chinese cooperation on energy and climate change: China would reap public health and agricultural benefits from reducing emissions, and the United States would earn goodwill for helping China do so.

By building on existing air-pollution agreements, the risk of distracting climate- change negotiations from the substantial task of promoting the reduction of carbon dioxide emissions could be avoided. Putting black carbon and ozone on the table in high-level climate talks could backfire if developing nations thought that they would be tacitly admitting responsibility for global warming by committing to reducing emissions of black carbon and ozone precursors or believed that the issue was an effort by developed countries to divert attention from the need for them to reduce their carbon dioxide emissions. Therefore, efforts to reduce emissions of black carbon and ozone precursors should be presented not as substitutes for commitments to reducing carbon dioxide emissions but as ways to quickly achieve local envi-ronmental and economic benefits.

The Low-hanging Fruit

Historically, initiatives to slow global warming have focused on reducing the emissions of carbon dioxide and other greenhouse gases and largely ignored the role played by air pollution. This strategy makes sense for the long run, since carbon dioxide emissions are, and will continue to be, the most important factor in climate change. But in the short run, it alone will not be enough. Some scientists have proposed geo-engineering-manipulating the climate through the use of technology- as a potential option of last resort, but the reduction of black carbon and ozone precursor emissions offers a less risky opportunity for achieving the same end.

Such an approach would quickly lower the level of black carbon and ozone in the atmosphere, offsetting the impact of decades of greenhouse gas emissions, decelerating the rush toward a dangerously warm planet, and giving efforts to reduce carbon dioxide emissions time to get off the ground. These pollutants are also tractable policy targets: they can be reduced through the use of existing technologies, institutions, and strategies, and doing so would lead to local improvements in air quality, agricultural output, and public health. In short, reducing black carbon and ozone precursor emissions is a low-risk, high-potential addition to the current arsenal of strategies to mitigate climate change.

At the current rate of global warming, the earth's temperature stands to careen out of control. Now is the time to look carefully at all the possible brakes that can be applied to slow climate change, hedge against near-term climate disasters, and buy time for technological innovations. Of the available strategies, focusing on reducing emissions of black carbon and ozone precursors is the low-hanging fruit: the costs are relatively low, the implementation is feasible, and the benefits would be numerous and immediate.

Appendix-I

FOURTH IPCC ASSESSMENT ON CLIMATE CHANGE*

SUMMARY FOR POLICYMAKERS: SOME EXTRACTS

With regard to changes in snow, ice and frozen ground there is high confidence that natural systems are affected. Examples are:

- enlargement and increased numbers of glacial lakes
- increasing ground instability and rock avalanches in mountain regions

Based on growing evidence, there is high confidence that the following effects on hydrological systems are occurring:

- increased runoff and earlier spring peak discharge in many glacier- and snow-fed rivers
- warming of lakes and rivers in many regions, with effects on thermal structure and water quality.

There is very high confidence, based on more evidence from a wider range of species, that recent warming is strongly affecting terrestrial biological systems, including such changes as:

---

* http://www.ipcc.ch/pdf/assessment-report/ar4/wg2/ar4-wg2-spm.pdf

- earlier timing of spring events, such as leaf-unfolding, bird migration and egg-laying

- poleward and upward shifts in ranges in plant and animal species

Based on satellite observations since the early 1980s, there is high confidence that there has been a trend towards earlier 'greening' of vegetation in the spring linked to longer thermal growing seasons due to recent warming.

There is high confidence, based on substantial new evidence, that observed changes in marine and freshwater biological systems are associated with rising water temperatures, as well as related changes in ice cover, salinity, oxygen levels and circulation. These include:

- shifts in ranges and changes in algal, plankton and fish abundance in high-latitude oceans

- increases in algal and zooplankton abundance in high-latitude and high-altitude lakes

- range changes and earlier migrations of fish in rivers

The uptake of anthropogenic carbon since 1750 has led to the ocean becoming more acidic. However, the effects of observed ocean acidification on the marine biosphere are as yet undocumented.

A global assessment of data since 1970 has shown it is likely that anthropogenic warming has had a discernible influence on many physical and biological systems.

Much more evidence has accumulated over the past five years to indicate that changes in many physical and biological systems are linked to anthropogenic warming. There are four sets of evidence which, taken together, support this conclusion:

1. The Working Group I Fourth Assessment concluded that most of the observed increase in the globally averaged temperature since the mid-20th century is very likely due to the observed increase in anthropogenic greenhouse gas concentrations.

2. Of the more than 29,000 observational data series, from 75 studies, that show significant change in many physical and biological systems, more than 89% are consistent with the direction of change expected as a response to warming.

3. A global synthesis of studies in this Assessment strongly demonstrates that the spatial agreement between regions of significant warming across the globe and the locations of significant observed changes in many systems consistent with warming is very unlikely to be due solely to natural variability of temperatures or natural variability of the systems.

4. Finally, there have been several modelling studies that have linked responses in some physical and biological systems to anthropogenic warming by comparing observed responses in these systems with modelled responses in which the natural forcings (solar activity and volcanoes) and anthropogenic forcings (greenhouse gases and aerosols) are explicitly separated. Models with combined natural and anthropogenic forcings simulate observed responses significantly better than models with natural forcing only. Limitations and gaps prevent more complete attribution of the causes of observed system responses to anthropogenic warming.

First, the available analyses are limited in the number of systems and locations considered. Second, natural temperature variability is larger at the regional than at the global scale, thus affecting identification of changes due to external forcing.

Finally, at the regional scale other factors (such as land-use change, pollution, and invasive species) are influential.

Nevertheless, the consistency between observed and modeled changes in several studies and the spatial agreement between significant regional warming and consistent impacts at the global scale is sufficient to conclude with high confidence that anthropogenic warming over the last three decades has had a discernible influence on many physical and biological systems.

Other effects of regional climate changes on natural and human environments are emerging, although many are difficult to discern due to adaptation and non-climatic drivers.

Recent climate changes and climate variations are beginning to have effects on many other natural and human systems.

However, based on the published literature, the impacts have not yet become established trends. Examples include:

- Settlements in mountain regions are at enhanced risk of glacier lake outburst floods caused by melting glaciers. Governmental institutions in some places have begun to respond by building dams and drainage works.
- In the Sahelian region of Africa, warmer and drier conditions have led to a reduced length of growing season with detrimental effects on crops. In southern Africa, longer dry seasons and more uncertain rainfall are prompting adaptation measures.
- Sea-level rise and human development are together contributing to losses of coastal wetlands and mangroves and increasing damage from coastal flooding in many areas.

Current knowledge about future impacts

The following is a selection of the key findings regarding projected impacts, as well as some findings on vulnerability and adaptation, in each system, sector and region for the range of (unmitigated) climate changes projected by the IPCC over this century judged to be relevant for people and the environment. The impacts frequently reflect projected changes in precipitation and other climate variables in addition to temperature, sea level and concentrations of atmospheric carbon dioxide. The magnitude and timing of impacts will vary with the amount and timing of climate change and, in some cases, the capacity to adapt.

More specific information is now available across a wide range of systems and sectors concerning the nature of future impacts, including for some fields not covered in previous assessments.

Freshwater resources and their management

By mid-century, annual average river runoff and water availability are projected to increase by 10-40% at high latitudes and in some wet tropical areas, and decrease by 10-30% over some dry regions at mid-latitudes and in the dry tropics, some of which are presently water-stressed areas.

Drought-affected areas will likely increase in extent. Heavy precipitation events, which are very likely to increase in frequency, will augment flood risk.

In the course of the century, water supplies stored in glaciers and snow cover are projected to decline, reducing water availability in regions supplied by melt water from major mountain ranges, where more than one-sixth of the world population currently lives.

Adaptation procedures and risk management practices for the water sector are being developed in some countries and regions that have recognised projected hydrological changes with related uncertainties.

Ecosystems

The resilience of many ecosystems is likely to be exceeded this century by an unprecedented combination of climate change, associated disturbances (e.g., flooding, drought, wildfire, insects, ocean acidification), and other global change drivers (e.g., land use change, pollution, over-exploitation of resources).

Over the course of this century, net carbon uptake by terrestrial ecosystems is likely to peak before mid-century and then weaken or even reverse, thus amplifying climate change.

Approximately 20-30% of plant and animal species assessed so far are likely to be at increased risk of extinction if increases in global average temperature exceed 1.5-2.5°C.

For increases in global average temperature exceeding 1.5-2.5°C and in concomitant atmospheric carbon dioxide concentrations, there are projected to be major changes in ecosystem structure and function, species' ecological interactions, and species' geographical ranges, with predominantly negative consequences for biodiversity, and ecosystem goods and services e.g., water and food supply.

The progressive acidification of oceans due to increasing atmospheric carbon dioxide is expected to have negative impacts on marine shell-forming organisms (e.g., corals) and their dependent species.

Food, fibre and forest products

Crop productivity is projected to increase slightly at mid- to high latitudes for local mean temperature increases of up to 1-3°C depending on the crop, and then decrease beyond that in some regions.

At lower latitudes, especially seasonally dry and tropical regions, crop productivity is projected to decrease for even small local temperature increases (1-2°C), which would increase the risk of hunger.

Globally, the potential for food production is projected to increase with increases in local average temperature over a range of 1-3°C, but above this it is projected to decrease.

Increases in the frequency of droughts and floods are projected to affect local crop production negatively, especially in subsistence sectors at low latitudes.

Adaptations such as altered cultivars and planting times allow low- and mid- to high-latitude cereal yields to be maintained at or above baseline yields for modest warming.

Globally, commercial timber productivity rises modestly with climate change in the short- to medium-term, with large regional variability around the global trend.

Regional changes in the distribution and production of particular fish species are expected due to continued warming, with adverse effects projected for aquaculture and fisheries.

Coastal systems and low-lying areas

Coasts are projected to be exposed to increasing risks, including coastal erosion, due to climate change and sea-level rise. The effect will be exacerbated by increasing human-induced pressures on coastal areas.

Corals are vulnerable to thermal stress and have low adaptive capacity. Increases in sea surface temperature of about 1-3°C are projected to result in more frequent coral bleaching events and widespread mortality, unless there is thermal adaptation or acclimatisation by corals.

Coastal wetlands including salt marshes and mangroves are projected to be negatively affected by sea-level rise especially where they are constrained on their landward side, or starved of sediment.

Many millions more people are projected to be flooded every year due to sea-level rise by the 2080s. Those densely-populated and low-lying areas where adaptive capacity is relatively low, and which already face other challenges such as tropical storms or local coastal subsidence, are especially at risk. The numbers affected will be largest in the mega-deltas of Asia while small islands are especially vulnerable.

Adaptation for coasts will be more challenging in developing countries than in developed countries, due to constraints on adaptive capacity.

Industry, settlement and society

Costs and benefits of climate change for industry, settlement and society will vary widely by location and scale. In the aggregate, however, net effects will tend to be more negative the larger the change in climate.

The most vulnerable industries, settlements and societies are generally those in coastal and river flood plains, those whose economies are closely linked with climate-sensitive resources, and those in areas prone to extreme weather events, especially where rapid urbanisation is occurring.

Poor communities can be especially vulnerable, in particular those concentrated in high-risk areas. They tend to have more limited adaptive capacities, and are more dependent on climate-sensitive resources such as local water and food supplies.

Where extreme weather events become more intense and/ or more frequent, the economic and social costs of those events will increase, and these increases will be substantial in the areas most directly affected. Climate change impacts spread from directly impacted areas and sectors to other areas and sectors through extensive and complex linkages.

Health

Projected climate change-related exposures are likely to affect the health status of millions of people, particularly those with low adaptive capacity, through:

- increases in malnutrition and consequent disorders, with implications for child growth and development;
- increased deaths, disease and injury due to heatwaves, floods, storms, fires and droughts;
- the increased burden of diarrhoeal disease;
- the increased frequency of cardio-respiratory diseases due to higher concentrations of ground-level ozone related to climate change; and,
- the altered spatial distribution of some infectious disease vectors.

The balance of positive and negative health impacts will vary from one location to another, and will alter over time as temperatures continue to rise. Critically important will be factors that directly shape the health of populations such as education, health care, public health initiatives and infrastructure and economic development.

More specific information is now available across the regions of the world concerning the nature of future impacts, including for some places not covered in previous assessments.

South and Southeast Asia

Glacier melt in the Himalayas is projected to increase flooding, and rock avalanches from destabilised slopes, and to affect water resources within the next two to three decades.

This will be followed by decreased river flows as the glaciers recede.

Freshwater availability in South and South-East Asia, particularly in large river basins, is projected to decrease due to climate change which, along with population growth and increasing demand arising from higher standards of living, could adversely affect more than a billion people by the 2050s.

Coastal areas, especially heavily-populated megadelta regions in South and South-East Asia, will be at greatest risk due to increased flooding from the sea and, in some megadeltas, flooding from the rivers.

Climate change is projected to impinge on the sustainable development of most developing countries of Asia, as it compounds the pressures on natural resources and the environment associated with rapid urbanisation, industrialisation, and economic development.

It is projected that crop yields could increase up to 20% in South-East Asia while they could decrease up to 30% in South Asia by the mid-21st century. Taken together, and considering the influence of rapid population growth and urbanisation, the risk of hunger is projected to remain very high in several developing countries.

Endemic morbidity and mortality due to diarrhoeal disease primarily associated with floods and droughts are expected to rise in South and South-East Asia due to projected changes in the hydrological cycle associated with global warming. Increases in coastal water temperature would exacerbate the abundance and/or toxicity of cholera in South Asia.

Magnitudes of impact can now be estimated more systematically for a range of possible increases in global average temperature.

Since the IPCC Third Assessment, many additional studies, particularly in regions that previously had been little researched, have enabled a more systematic understanding of how the timing and magnitude of impacts may be affected by changes in climate and sea level associated with differing amounts and rates of change in global average temperature.

Examples of this new information are presented in Figure SPM.2. Entries have been selected which are judged to be relevant for people and the environment and for which there is high confidence in the assessment. All examples of impact are drawn from chapters of the Assessment, where more detailed information is available.

Depending on circumstances, some of these impacts could be associated with 'key vulnerabilities', based on a number of criteria in the literature (magnitude, timing, persistence/reversibility, the potential for adaptation, distributional aspects, likelihood and 'importance' of the impacts). Assessment of potential key vulnerabilities is intended to provide information on rates and levels of climate change to help decision-makers make appropriate responses to the risks of climate change.

Illustrative examples of global impacts projected for climate changes (and sea level and atmospheric carbon dioxide where relevant) associated with different amounts of increase in global average surface temperature in the 21st century. The black lines link impacts, dotted arrows indicate impacts continuing with increasing temperature. Entries are placed so that the left-hand side of the text indicates the approximate onset of a given impact. Quantitative entries for water stress and flooding represent the additional impacts of climate change relative to the conditions projected across the range of Special Report on Emissions Scenarios (SRES). Adaptation to climate change is not included in these estimations. All entries are from published studies recorded in the chapters of the Assessment. Confidence levels for all statements are high.

Key impacts as a function of increasing global average temperature change

(Impacts will vary by extent of adaptation, rate of temperature change, and socio-economic pathway)

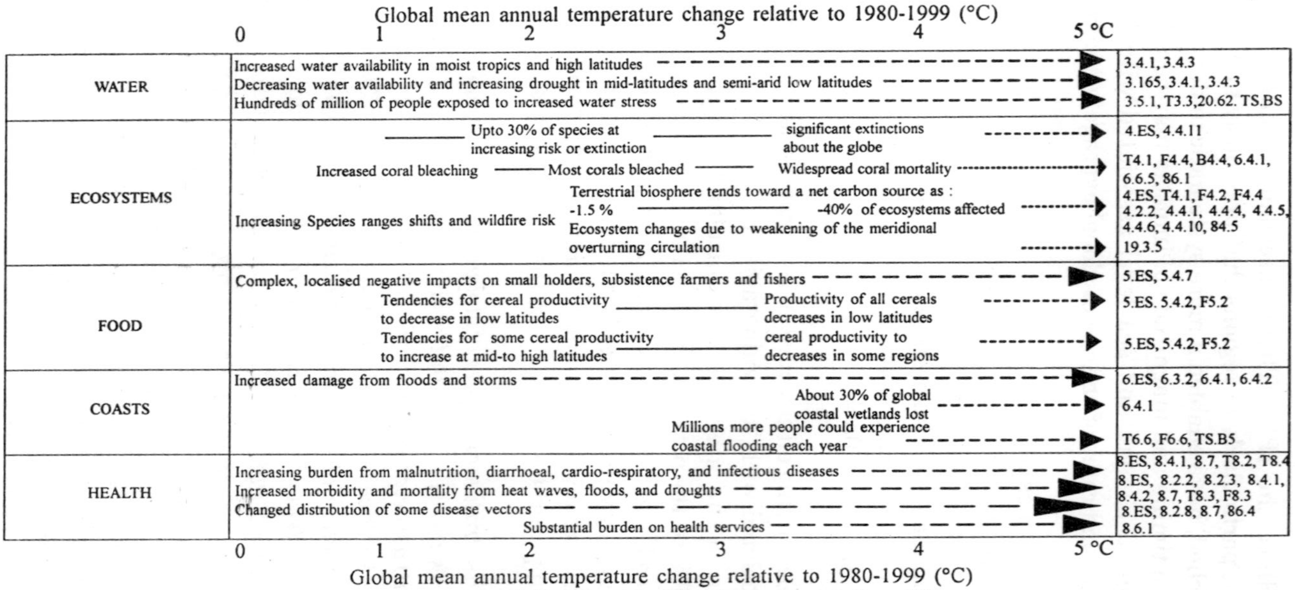

* Significant is defined here as more than 40%

* Based on average rate of sea level rise of 4.2 min/year from 2000 to 2080

Fig. SPM.2

Impacts due to altered frequencies and intensities of extreme weather, climate and sea-level events are very likely to change.

Since the IPCC Third Assessment, confidence has increased that some weather events and extremes will become more frequent, more widespread and/or more intense during the 21st century; and more is known about the potential effects of such changes. A selection of these is presented in Table SPM.1.

Some large-scale climate events have the potential to cause very large impacts, especially after the 21st century.

Very large sea-level rises that would result from widespread deglaciation of Greenland and West Antarctic ice sheets imply major changes in coastlines and ecosystems, and inundation of low-lying areas, with greatest effects in river deltas. Relocating populations, economic activity, and infrastructure would be costly and challenging. There is medium confidence that at least partial deglaciation of the Greenland ice sheet, and possibly the West Antarctic ice sheet, would occur over a period of time ranging from centuries to millennia for a global average temperature increase of 1-4°C (relative to 1990-2000), causing a contribution to sea-level rise of 4-6 m or more. The complete melting of the Greenland ice sheet and the West Antarctic ice sheet would lead to a contribution to sea-level rise of up to 7 m and about 5 m, respectively

Impacts of climate change will vary regionally but, aggregated and discounted to the present, they are very likely to impose net annual costs which will increase over time as global temperatures increase.

This Assessment makes it clear that the impacts of future climate change will be mixed across regions. For increases in global mean temperature of less than 1-3°C above 1990 levels, some impacts are projected to produce benefits in some places and some sectors, and produce costs in other places and other sectors. It is, however, projected that some low-latitude regions like south and southeast Asia will experience net costs even for small increases in temperature.

Many estimates of aggregate net economic costs of damages from climate change across the globe (i.e., the social cost of carbon (SCC), expressed in terms of future net benefits and costs that are discounted to the present) are now available. Peer-reviewed estimates of the SCC for 2005 have an average value of US$43 per tonne of carbon (i.e., US$12 per tonne of carbon dioxide), but the range around this mean is large. For example, in a survey of 100 estimates, the values ran from US$-10 per tonne of carbon (US$-3 per tonne of carbon dioxide) up to US$350 per tonne of carbon (US$95 per tonne of carbon dioxide).

The large ranges of SCC are due in the large part to differences in assumptions regarding climate sensitivity, response lags, the treatment of risk and equity, economic and non-economic impacts, the inclusion of potentially catastrophic losses, and discount rates. It is very likely that globally aggregated figures underestimate the damage costs because they cannot include many non-quantifiable impacts. Taken as a whole, the range of published evidence indicates that the net damage costs of climate change are likely to be significant and to increase over time.

It is virtually certain that aggregate estimates of costs mask significant differences in impacts across sectors, regions, countries and populations. In some locations and among some groups of people with high exposure, high sensitivity and/or low adaptive capacity, net costs will be significantly larger than the global aggregate.

Phenomenon and direction of trend	*Likehood of future trends based on projection for 21st century using SRES scenarios*	*Examples of major projected impacts by sector*			
		Agriculture, forestry and ecosystems [4.4, 5.4]	*Water resources [3.4]*	*Human health [8.2, 8.4]*	*Industry settlement and society [7.4]*
Over most land areas, warmer and fewer cold days and nights, warmer and more frequent hot days and nights	Virtually certain	Increased yield in colder environments, decreased yields in warmer environments; increased insect outbreaks	Effects on water resources relying on snow melt; effects on some water supplies	Reduced human mortality from decreased cold exposure	Reduced enery demand for heating; increased demand for cooling; declining air quality in cities; reduced disruption to transport due to snow, ice; effects on winter tourism
Warm spells/heat waves, Frequency increases over most land areas	Very likely	Reduced yields in warmer regions due to heat stress; increased danger of wildfire	Increased water demand; water quality problem, e.g., algal blooms	Increased risk of heat related mortality, especially for the elderly chronically sick, very young and socially-isolated	Reduction in quality of life for people in warm areas without appropriate housing impacts on the elderly, very young and poor
Heavy precipitation events, Frequency increases over most areas	Very likely	Damage to crops; soil erosion, inability to cultivate land due to waterlogging of soils	Adverse effects on quality of surface and groundwater; contarmination of water supply; water scarity may be relieved	Increased risk of death, injuries and infectious, respiratory and skin diseases	Disruption of settlements, commerce, transport and societies due to flooding; pressures on urban and rural infrastructures; loss of property
Area affected by drought increases	Likely	Land degradation; lower yields/crop damage and failure; increased livestock deaths; increased risk of wildfire	More widespread water stress	Increased risk of food and water shortage; increased risk of malnutrition; increased risk of water and food borne diseases	Water shortages for settlements, industry and societies, reduced hydropower generation potentials; potential for population migration
Intense tropical cyclone activity increases	Likely	Damage to crops; windthrow (uprooting) of trees; damage to corel reefs	Power outages causing disruption of public water supply	Increased risk of deaths, injuries, water and food-borne diseases; post-traumatic stress disorders	Disruption by flood and high winds, withdrawal of risk coverage in vulnerable areas by private insurers, potential for population migration, loss of property
Increased incidence of extreme high sea level (excludes taunamia)	Likely	Salinisation of irrigation water, estuaries and freshwater systems	Decreased freshwater availability due to saltwater intrusion	Increased risk of deaths and injuries by drowing in floods; migration-related health effects	Costs of coastal protection versus costs of land-use relocation; potential for movement of populations and infrastructure; also see tropical cyclones above

a See Working Group I Fouth Assessment Table 3.7 for further details regarding definitions.

b Warming of the most extreme days and nights each year.

c Extreme high sea level depends on average sea level and on regional weather systems. It is declined as the highest 1% of hourly values of observed sea level at a station for a given reference period.

d In all scenarios, the projected global average sea level at 2100 is higher than in the reference period [Working Group I Fourth Assessment 10.6]. The effect of changes in regional weather systems on sea level extremes has not been assessed.

Table SPM.1

Examples of possible impacts of climate change due to changes in extreme weather and climate events, based on projections to the mid- to late 21st century. These do not take into account any changes or developments in adaptive capacity. Examples of all entries are to be found in chapters in the full Assessment (see source at top of columns). The first two columns of the table (shaded yellow) are taken directly from the Working Group I Fourth Assessment (Table SPM-2). The likelihood estimates in Column 2 relate to the phenomena listed in Column 1.

Current knowledge about responding to climate change

Some adaptation is occurring now, to observed and projected future climate change, but on a limited basis.

There is growing evidence since the IPCC Third Assessment of human activity to adapt to observed and anticipated climate change. For example, climate change is considered in the design of infrastructure projects such as coastal defence in the Maldives Other examples include prevention of glacial lake outburst flooding in Nepal, and policies and strategies such as water management in Australia

Adaptation will be necessary to address impacts resulting from the warming which is already unavoidable due to past emissions.

Past emissions are estimated to involve some unavoidable warming (about a further 0.6°C by the end of the century relative to 1980-1999) even if atmospheric greenhouse gas concentrations remain at 2000 levels. There are some impacts for which adaptation is the only available and appropriate response. An indication of these impacts can be seen in Figure SPM.2.

A wide array of adaptation options is available, but more extensive adaptation than is currently occurring is required to reduce vulnerability to future climate change. There are barriers, limits and costs, but these are not fully understood.

Impacts are expected to increase with increases in global average temperature, as indicated in Figure SPM. 2. Although many early impacts of climate change can be effectively addressed through adaptation, the options for successful adaptation diminish and the associated costs increase with increasing climate change. At present we do not have a clear picture of the limits to adaptation, or the cost, partly because effective adaptation measures are highly dependent on specific, geographical and climate risk factors as well as institutional, political and financial constraints.

The array of potential adaptive responses available to human societies is very large, ranging from purely technological (e.g., sea defences), through behavioural (e.g., altered food and recreational choices), to managerial (e.g., altered farm practices) and to policy (e.g., planning regulations). While most technologies and strategies are known and developed in some countries, the assessed literature does not indicate how effective various options are at fully reducing risks, particularly at higher levels of warming and related impacts, and for vulnerable groups. In addition, there are formidable environmental, economic, informational, social, attitudinal and behavioural barriers to the implementation of adaptation. For developing countries, availability of resources and building adaptive capacity are particularly important.

Adaptation alone is not expected to cope with all the projected effects of climate change, and especially not over the long term as most impacts increase in magnitude.

Vulnerability to climate change can be exacerbated by the presence of other stresses.

Non-climate stresses can increase vulnerability to climate change by reducing resilience and can also reduce adaptive capacity because of resource deployment to competing needs. For example, current stresses on some coral reefs include marine pollution and chemical runoff from agriculture as well as increases in water temperature and ocean acidification. Vulnerable regions face multiple stresses that affect their exposure and sensitivity as well as their capacity to adapt. These stresses arise from, for example, current climate hazards, poverty and unequal access to resources, food insecurity, trends in economic globalisation, conflict, and incidence of diseases such as HIV/AIDS. Adaptation measures are seldom undertaken in response to climate change alone but can be integrated within, for example, water resource management, coastal defence and risk-reduction strategies.

Future vulnerability depends not only on climate change but also on development pathway.

The IPCC studies show that the projected impacts of climate change can vary greatly due to the development pathway assumed. For example, there may be large differences in regional population, income and technological development under alternative scenarios, which are often a strong determinant of the level of vulnerability to climate change.

To illustrate, in a number of recent studies of global impacts of climate change on food supply, risk of coastal flooding and water scarcity, the projected number of people affected is considerably greater under the scenario of development (characterised by relatively low per capita income and large population growth) than under other futures. This difference is largely explained, not by differences in changes of climate, but by differences in vulnerability.

Sustainable development can reduce vulnerability to climate change, and climate change could impede nations' abilities to achieve sustainable development pathways.

Sustainable development can reduce vulnerability to climate change by enhancing adaptive capacity and increasing resilience. At present, however, few plans for promoting sustainability have explicitly included either adapting to climate change impacts, or promoting adaptive capacity.

On the other hand, it is very likely that climate change can slow the pace of progress towards sustainable development, either directly through increased exposure to adverse impact or indirectly through erosion of the capacity to adapt.

Many impacts can be avoided, reduced or delayed by mitigation.

A small number of impact assessments have now been completed for scenarios in which future atmospheric concentrations of greenhouse gases are stabilised. Although these studies do not take full account of uncertainties in projected climate under stabilisation, they nevertheless provide indications of damages avoided or vulnerabilities and risks reduced for different amounts of emissions reduction.

A portfolio of adaptation and mitigation measures can diminish the risks associated with climate change.

Even the most stringent mitigation efforts cannot avoid further impacts of climate change in the next few decades, which makes adaptation essential, particularly in addressing near-term impacts. Unmitigated climate change would, in the long term, be likely to exceed the capacity of natural, managed and human systems to adapt.

This suggests the value of a portfolio or mix of strategies that includes mitigation, adaptation, technological development (to enhance both adaptation and mitigation) and research (on climate science, impacts, adaptation and mitigation). Such portfolios could combine policies with incentive-based approaches, and actions at all levels from the individual citizen through to national governments and international organisations.

One way of increasing adaptive capacity is by introducing the consideration of climate change impacts in development planning for example, by:

- including adaptation measures in land-use planning and infrastructure design
- including measures to reduce vulnerability in existing disaster risk reduction strategies

Definitions of key terms

Climate change in IPCC usage refers to any change in climate over time, whether due to natural variability or as a result of human activity. This usage differs from that in the Framework Convention on Climate Change, where climate change refers to a change of climate that is attributed directly or indirectly to human activity that alters the composition of the global atmosphere and that is in addition to natural climate variability observed over comparable time periods.

Adaptive capacity is the ability of a system to adjust to climate change (including climate variability and extremes) to moderate potential damages, to take advantage of opportunities, or to cope with the consequences.

Vulnerability is the degree to which a system is susceptible to, and unable to cope with, adverse effects of climate change, including climate variability and extremes. Vulnerability is a function of the character, magnitude, and rate of climate change and variation to which a system is exposed, its sensitivity, and its adaptive capacity.

Communication of Uncertainty in Fourth Assessment

Terminology	**Degree of confidence in being correct**
Very high confidence	At least 9 out of 10 chance of being correct
High confidence	About 8 out of 10 chance
Medium confidence	About 5 out of 10 chance
Low confidence	About 2 out of 10 chance
Very low confidence	Less than a 1 out of 10 chance

Terminology	**Likelihood of the occurrence/ outcome**
Virtually certain	>99% probability of occurrence
Very likely	90 to 99% probability
Likely	66 to 90% probability
About as likely as not	33 to 66% probability
Unlikely	10 to 33% probability
Very unlikely	1 to 10% probability
Exceptionally unlikely	<1% probability

Appendix-II

Extracts from Human Development Report 2007-2008*

Fighting climate change: human solidarity in a divided world

According to Martin Luther King Jr., *"Human progress is neither automatic nor inevitable. We are faced now with the fact that tomorrow is today. We are confronted with the fierce urgency of now. In this unfolding conundrum of life and history there is such a thing as being too late...We may cry out desperately for time to pause in her passage, but time is deaf to every plea and rushes on. Over the bleached bones and jumbled residues of numerous civilizations are written the pathetic words: Too late."*

Delivered in a sermon on social justice four decades ago, Martin Luther King's words retain a powerful resonance. At the start of the 21st Century, we too are confronted with the "fierce urgency" of a crisis that links today and tomorrow. That crisis is climate change. It is still a preventable crisis—but only just. The world has less than a decade to change course. No issue merits more urgent attention—or more immediate action.

Climate change is the defining human development issue of our generation. All development is ultimately about expanding human potential and enlarging human freedom. It is about people developing the capabilities that empower them to make

* http://hdr.undp.org/en/reports/global/her2007-2008.

choices and to lead lives that they value. Climate change threatens to erode human freedoms and limit choice. It calls into question the Enlightenment principle that human progress will make the future look better than the past.

The early warning signs are already visible. Today, we are witnessing at first hand what could be the onset of major human development reversal in our lifetime. Across developing countries, millions of the world's poorest people are already being forced to cope with the impacts of climate change. These impacts do not register as apocalyptic events in the full glare of world media attention. They go unnoticed in financial markets and in the measurement of world gross domestic product (GDP).

How the world deals with climate change today will have a direct bearing on the human development prospects of a large section of humanity. Failure will consign the poorest 40 percent of the world's population—some 2.6 billion people—to a future of diminished opportunity. It will exacerbate deep inequalities within countries. And it will undermine efforts to build a more inclusive pattern of globalization, reinforcing the vast disparities between the 'haves' and the 'have nots'.

The rapid build-up of greenhouse gases in the Earth's atmosphere is fundamentally changing the climate forecast for future generations. We are edging towards 'tipping points'. These are unpredictable and non-linear events that could open the door to ecological catastrophes—accelerated collapse of the Earth's great ice sheets being a case in point—that will transform patterns of human settlement and undermine the viability of national economies. Aversion to poverty and inequality today, and to catastrophic risk in the future provides a strong rationale for urgent action.

Some commentators continue to cite uncertainty over future outcomes as grounds for a limited response to climate change. That starting point is flawed. There are indeed many unknowns: climate science deals in probability and risk, not in certainties. However, if we value the well-being of our children and grandchildren, even small risks of catastrophic events merit an insurance-based precautionary approach. And uncertainty cuts both ways: the risks could be greater than we currently understand.

It raises profoundly important questions about social justice, equity and human rights across countries and generations. In the Human Development Report 2007/2008 we address these questions. Our starting point is that the battle against climate change can—and must—be won. The world lacks neither the financial resources nor the technological capabilities to act. If we fail to prevent climate change it will be because we were unable to foster the political will to cooperate.

Such an outcome would represent not just a failure of political imagination and leadership, but a moral failure on a scale unparalleled in history. During the 20th Century failures of political leadership led to two world wars. Millions of people paid a high price for what were avoidable catastrophes. Dangerous climate change is the avoidable catastrophe of the 21st Century and beyond. Future generations will pass a harsh judgement on a generation that looked at the evidence on climate change, understood the consequences and then continued on a path that consigned millions of the world's most vulnerable people to poverty and exposed future generations to the risk of ecological disaster.

Ecological interdependence

Climate change is different from other problems facing humanity—and it challenges us to think differently at many levels. Above all, it challenges us to think about what it means to live as part of an ecologically interdependent human community.

Ecological interdependence is not an abstract concept. We live today in a world that is divided at many levels. People are separated by vast gulfs in wealth and opportunity. In many regions, rival nationalisms are a source of conflict. All too often, religious, cultural and ethnic identity are treated as a source of division and difference from others. In the face of all these differences, climate change provides a potent reminder of the one thing that we share in common. It is called planet Earth. All nations and all people share the same atmosphere. And we only have one.

Global warming is evidence that we are overloading the carrying capacity of the Earth's atmosphere. Stocks of greenhouse gases that trap heat in the atmosphere are accumulating at an unprecedented rate. Current concentrations have reached 380 parts per million (ppm) of carbon dioxide equivalent (CO_2e) exceeding the natural range of the last 650,000 years. In the course of the 21st Century, average global temperatures could increase by more than 5°C.

To put that figure in context, it is equivalent to the change in temperature since the last ice age — an era in which much of Europe and North America was under more than one kilometre of ice. The threshold for dangerous climate change is an increase of around 2°C. This threshold broadly defines the point at which rapid reversals in human development and a drift towards irreversible ecological damage would become very difficult to avoid.

Behind the numbers and the measurement is a simple overwhelming fact. We are recklessly mismanaging our ecological interdependence. In effect, our generation is running up an unsustainable ecological debt that future generations will inherit. We are drawing down the stock of environmental capital of our children. Dangerous climate change will represent the adjustment to an unsustainable level of greenhouse gas emissions.

Future generations are not the only constituency that will have to cope with a problem they did not create. The world's poor will suffer the earliest and most damaging impacts. Rich nations and their citizens account for the overwhelming bulk of the greenhouse gases locked in the Earth's atmosphere. But, poor countries and their citizens will pay the highest price for climate change.

The inverse relationship between responsibility for climate change and vulnerability to its impacts is sometimes forgotten. Public debate in rich nations increasingly highlights the threat posed by rising greenhouse gas emissions from developing countries. That threat is real. But it should not obscure the underlying problem. Mahatma Gandhi once reflected on how many planets might be needed if India were to follow Britain's pattern of industrialization. We are unable to answer that question. However, we estimate in this Report that if all of the world's people generated greenhouse gases at the same rate as some developed countries, we would need nine planets.

While the world's poor walk the Earth with a light carbon footprint they are bearing the brunt of unsustainable management of our ecological interdependence. In rich countries, coping with climate change to date has largely been a matter of adjusting thermostats, dealing with longer, hotter summers, and observing seasonal shifts. Cities like London and Los Angeles may face flooding risks as sea levels rise, but their inhabitants are protected by elaborate flood defence systems. By contrast, when global warming changes weather patterns in the Horn of Africa, it means that crops fail and people go hungry, or that women and young girls spend more hours collecting water. And, whatever the future risks facing cities in the rich world, today the real climate change vulnerabilities linked to storms and floods are to be found in rural communities in the great river deltas of the Ganges, the Mekong and the Nile, and in sprawling urban slums across the developing world.

The emerging risks and vulnerabilities associated with climate change are the outcomes of physical processes. But they are also a consequence of human actions and choices. This is another aspect of ecological interdependence that is sometimes forgotten. When people in an American city turn on the airconditioning or people in Europe drive their cars, their actions have consequences. Those consequences link them to rural communities in Bangladesh, farmers in Ethiopia and slum dwellers in Haiti. With these human connections come moral responsibilities, including a responsibility to reflect upon—and change—energy policies that inflict harm on other people or future generations.

The case for action

If the world acts now it will be possible—just possible—to keep 21st Century global temperature increases within a 2°C threshold above preindustrial levels. Achieving this future will require a high level of leadership and unparalleled international cooperation. Yet climate change is a threat that comes with an opportunity. Above all, it provides an opportunity for the world to come together in forging a collective response to a crisis that threatens to halt progress.

The values that inspired the drafters of the Universal Declaration of Human Rights provide a powerful point of reference. That document was a response to the political failure that gave rise to extreme nationalism, fascism and world war. It established a set of entitlements and rights—civil, political, cultural, social and economic — for "all members of the human family". The values that inspired the Universal Declaration were seen as a code of conduct for human affairs that would prevent the "disregard and contempt for human rights that have resulted in barbarous acts which have outraged the conscience of mankind".

The drafters of the Universal Declaration of Human Rights were looking back at a human tragedy, the second world war, that had already happened. Climate change is different. It is a human tragedy in the making. Allowing that tragedy to evolve would be a political failure that merits the description of an "outrage to the conscience of mankind". It would represent a systematic violation of the human rights of the world's poor and future generations and a step back from universal values. Conversely, preventing dangerous climate change would hold out the hope for the development of multilateral solutions to the wider problems facing the international community. Climate change confronts us with enormously complex questions that span science, economics and international relations. These questions have to be addressed through practical strategies. Yet it is important not to lose sight of the wider issues that are at stake. The real choice facing political leaders and people today is between universal human values, on the one side, and participating in the widespread and systematic violation of human rights on the other.

The starting point for avoiding dangerous climate change is recognition of three distinctive features of the problem. The first feature is the combined force of inertia and cumulative outcomes of climate change. Once emitted, carbon dioxide (CO_2) and other greenhouse gases stay in the atmosphere for a long time. There are no rapid rewind buttons for running down stocks. People living at the start of the 22nd Century will live with the consequences of our emissions, just as we are living with the consequences of emissions since the industrial revolution. Time-lags are an important consequence of climate change inertia. Even stringent mitigation measures will not materially affect average temperatures changes until the mid-2030s—and temperatures will not peak until 2050. In other words, for the first half of the 21st Century the world in general, and the world's poor in particular, will have to live with climate change to which we are already committed.

The cumulative nature of the climate change has wide-ranging implications. Perhaps the most important is that carbon cycles do not follow political cycles. The current generation of political leaders cannot solve the climate change problem alone because a sustainable emissions pathway has to be followed over decades, not years. However, it has the power either to prise open the window of opportunity for future generations, or to close that window. Urgency is the second feature of the climate change challenge—and a corollary of inertia. In many other areas of international relations, inaction or delayed agreements have limited costs. International trade is an example. This is an area in which negotiations can break down and resume without inflicting long-term damage on the underlying system—as witnessed by the unhappy history of the Doha Round. With climate change, every year of delay in reaching an agreement to cut emissions adds to greenhouse gas stocks, locking the future into a higher temperature. In the seven years since the Doha Round started, to continue the analogy, stocks of greenhouse gases have increased by around 12 ppm of CO_2e—and those stocks will still be there when the trade rounds of the 22nd Century get underway.

There are no obvious historical analogies for the urgency of the climate change problem. During the Cold War, large stockpiles of nuclear missiles pointed at cities posed a grave threat to human security. However, 'doing nothing' was a strategy for containment of the risks. Shared recognition of the reality of mutually assured destruction offered a perversely predictable stability. With climate change, by contrast, doing nothing offers a guaranteed route to a further build-up greenhouse gases, and to mutually assured destruction of human development potential.

The third important dimension of the climate change challenge is its global scale. The Earth's atmosphere does not differentiate greenhouse gases by country of origin. One tonne of greenhouse gases from China carries the same weight as one tonne of greenhouse gases from the United States—and one country's emissions are another country's climate change problem. It follows that no one country can win the battle against climate change acting alone. Collective action is not an option but an imperative.

Seizing the moment—2012 and beyond

Confronted with a problem as daunting as climate change, resigned pessimism might seem a justified response. However, resigned pessimism is a luxury that the world's poor and future generations cannot afford—and there is an alternative.

There is cause for optimism. Five years ago, the world was still engaged in debating whether or not climate change was taking place, and whether or not it was human-induced. Climate change scepticism was a flourishing industry. Today, the debate is over and climate scepticism is an increasingly fringe activity. The fourth assessment review of the International Panel on Climate Change has established an overwhelming scientific consensus that climate change is both real and man-made.

Political momentum is also gathering pace. Many governments are setting bold targets for cutting greenhouse gas emissions. Climate change mitigation has now registered firmly on the agenda of the Group of Eight (G8) industrialized nations. And dialogue between developed and developing countries is strengthening.

All of this is positive news. Practical outcomes are less impressive. While governments may recognize the realities of global warming, political action continues to fall far short of the minimum needed to resolve the climate change problem. The gap between scientific evidence and political response remains large. In the developed world, some countries have yet to establish ambitious targets for cutting greenhouse gas emissions. Others have set ambitious targets without putting in place the energy policy reforms needed to achieve them. The deeper problem is that the world lacks a clear, credible and long-term multilateral framework that charts a course for avoiding dangerous climate change—a course that spans the divide between political cycles and carbon cycles.

With the expiry of the current commitment period of the Kyoto Protocol in 2012, the international community has an opportunity to put that framework in place. Seizing that opportunity will require bold leadership. Missing it will push the world further on the route to dangerous climate change.

Developed countries have to take the lead. They carry the burden of historic responsibility for the climate change problem. And they have the financial resources and technological capabilities to initiate deep and early cuts in emissions. Putting a price on carbon through taxation or cap-and-trade systems is the starting point. But market pricing alone will not be enough. The development of regulatory systems and public–private partnerships for a low-carbon transition are also priorities.

The principle of "common but differentiated responsibility"—one of the foundations of the Kyoto framework—does not mean that developing countries should do nothing. The credibility of any multilateral agreement will hinge on the participation of major emitters in the developing world. However, basic principles of equity and the human development imperative of expanding access to energy demand that developing countries have the flexibility to make the transition to a low-carbon growth path at a rate consistent with their capabilities.

International cooperation has a critical role to play at many levels. The global mitigation effort would be dramatically enhanced if a post-2012 Kyoto framework incorporated mechanisms for finance and technology transfers. These mechanisms could help remove obstacles to the rapid disbursement of the lowcarbon technologies needed to avoid dangerous climate change.

Adaptation priorities must also be addressed. For too long, climate change adaptation has been treated as a peripheral concern, rather than as a core part of the international poverty reduction agenda. Mitigation is an imperative because it will define prospects for avoiding dangerous climate change in the future. But the world's poor cannot be left to sink or swim with their own resources while rich countries protect their citizens behind climate-defence fortifications. Social justice and respect of human rights demand stronger international commitment on adaptation.

Our legacy

The post-2012 Kyoto framework will powerfully influence prospects for avoiding climate change—and for coping with the climate change that is now unavoidable. Negotiations on that framework will be shaped by governments with very different levels of negotiating leverage. Powerful vested interests in the corporate sector will also make their voices heard. As governments embark on the negotiations for a post-2012 Kyoto Protocol, it is important that they reflect on two constituencies with a limited voice but a powerful claim to social justice and respect for human rights: the world's poor and future generations.

People engaged in a daily struggle to improve their lives in the face of grinding poverty and hunger ought to have first call on human solidarity. They certainly deserve something more than political leaders who gather at international summits, set high-sounding development targets and then undermine achievement of the very same targets by failing to act on climate change. And our children and their children's grandchildren have the right to hold us to a high standard of accountability when their future—and maybe their survival—is hanging in the balance. They too deserve something more than a generation of political leaders who look at the greatest challenge humankind has ever faced and then sit on their hands.

The battle against dangerous climate change is part of the fight for humanity. Winning that battle will require far-reaching changes at many levels—in consumption, in how we produce and price energy, and in international cooperation. Above all, though, it will require far-reaching changes in how we think about our ecological interdependence, about social justice for the world's poor, and about the human rights and entitlements of future generations.

The 21st Century climate challenge

Global warming is already happening. World temperatures have increased by around 0.7°C since the advent of the industrial era—and the rate of increase is quickening. There is overwhelming scientific evidence linking the rise in temperature to increases in the concentration of greenhouse gases in the Earth's atmosphere.

There is no hard-and-fast line separating 'dangerous' from 'safe' climate change. Many of the world's poorest people and most fragile ecological systems are already being forced to adapt to dangerous climate change. However, beyond a threshold of 2°C the risk of large-scale human development setbacks and irreversible ecological catastrophes will increase sharply.

Business-as-usual trajectories will take the world well beyond that threshold. To have a 50:50 chance of limiting temperature increase to 2°C above preindustrial levels will require stabilization of greenhouse gases.

Temperature scenarios do not capture the potential human development impacts. Average changes in temperature on the scale projected in business-as-usual scenarions will trigger large-scale reversals in human development, undermining livelihoods and causing mass displacement. By the end of the 21st Century, the spectre of catastrophic ecological impacts could have moved from the bounds of the possible to the probable.

Countries vary widely in their contribution to the emissions that are driving up atmospheric stocks of greenhouse gases. With 15 percent of world population, rich countries account for almost half of emissions of CO_2. High growth in China and India is leading to a gradual convergence in 'aggregate' emissions. However, per capita carbon footprint convergence is more limited. The carbon footprint of the United States is five times that of China and over 15 times that of India.

Our stabilization target is stringent but affordable. Between now and 2030, the average annual cost would amount to 1.6 percent of GDP. This is not an insignificant investment. But it represents less than two-thirds of global military spending. The costs of inaction could be much higher.

Current investment patterns are putting in place a carbon intensive energy infrastructure, with coal playing a dominant role. On the basis of current trends and present policies, energy-related CO_2 emissions could rise by more than 50 percent over 2005 levels by 2030.

Climate shocks: risk and vulnerability in an unequal world

Climate shocks already figure prominently in the lives of the poor. Events such as droughts, fl oods and storms are often terrible experiences for those affected: they threaten lives and leave people feeling insecure. But climate shocks also erode long-term opportunities for human development, undermining productivity and eroding human capabilities. No single climate shock can be attributed to climate change. However, climate change is ratcheting up the risks and vulnerabilities facing the poor. It is placing further stress on already over-stretched coping mechanisms and trapping people in downward spirals of deprivation.

High levels of poverty and low levels of human development limit the capacity of poor households to manage climate risks. With limited access to formal insurance, low incomes and meagre assets, poor households have to deal with climate-related shocks under highly constrained conditions.

Strategies for coping with climate risks can reinforce deprivation. Producers in drought prone areas often forego production of crops that could raise income in order to minimize risk, preferring to produce crops with lower economic returns but resistant to drought. When climate disasters strike, the poor are often forced to sell productive assets, with attendant implications for recovery, in order to protect consumption. And when that is not enough households cope in other ways: for example, by cutting meals, reducing spending on health and taking children out of school. These are desperation measures that can create life-long cycles of disadvantage, locking vulnerable households into low human development traps.

Research carried out for this report underlines just how potent these traps can be. Using microlevel household data we examined some of the long-term impacts of climate-shocks in the lives of the poor. In Ethiopia and Kenya, two of the world's most drought-prone countries, children aged five or less are respectively 36 and 50 percent more likely to be malnourished if they were born during a drought. For Ethiopia, that translates into some 2 million additional malnourished children in 2005. In Niger, children aged two or less born in a drought year were 72 percent more likely to be stunted. And Indian women born during a flood in the 1970s were 19 percent less likely to have attended primary school.

The long-run damage to human development generated through climate shocks is insufficiently appreciated. Media reporting of climate-related disasters often plays an important role in informing opinion—and in capturing the human suffering that comes with climate shocks. However, it also gives rise to a perception that these are 'here-today-gone-tomorrow' experiences, diverting attention from the long-run human consequences of droughts and floods.

Climate change will not announce itself as an apocalyptic event in the lives of the poor. Direct attribution of any specific event to climate change will remain impossible. However, climate change will steadily increase the exposure of poor and vulnerable households to climate-shocks and place increased pressure on coping strategies, which, over time, could steadily erode human capabilities.

We identify five key transmission mechanisms through which climate change could stall and then reverse human development:

- Agricultural production and food security. Climate change will affect rainfall, temperature and water availability for agriculture in vulnerable areas. For example, drought affected areas in sub-Saharan Africa could expand by 60–90 million hectares, with dry land zones suffering losses of US$26 billion by 2060 (2003 prices), a figure in excess of bilateral aid to the region in 2005. Other developing regions—including Latin America and South Asia—will also experience losses in agricultural production, undermining efforts to cut rural poverty. The additional number affected by malnutrition could rise to 600 million by 2080.

- Water stress and water insecurity. Changed run-off patterns and glacial melt will add to ecological stress, compromising flows of water for irrigation and human settlements in the process. An additional 1.8 billion people could be living in a water scarce environment by 2080. Central Asia, Northern China and the northern part of South Asia face immense vulnerabilities associated with the retreat of glaciers—at a rate of 10–15 metres a year in the Himalayas. Seven of Asia's great river systems will experience an increase in flows over the short term, followed by a decline as glaciers melt. The Andean region also faces imminent water security threats with the collapse of tropical glaciers. Several countries in already highly water-stressed regions such as the Middle East could experience deep losses in water availability.

- Rising sea levels and exposure to climate disasters. Sea levels could rise rapidly with accelerated ice sheet disintegration. Global temperature increases of 3–4°C could result in 330 million people being permanently or temporarily displaced through flooding. Over 70 million people in Bangladesh, 6 million in Lower Egypt and 22 million in Viet Nam could be affected. Small island states in the Caribbean and Pacific could suffer catastrophic damage. Warming seas will also fuel more intense tropical storms. With over 344 million people currently exposed to tropical cyclones, more intensive storms could have devastating consequences for a large group of countries. The 1 billion people currently living in urban slums on fragile hillsides or flood-prone river banks face acute vulnerabilities.

- Ecosystems and biodiversity. Climate change is already transforming ecological systems. Around one-half of the world's coral reef systems have suffered 'bleaching' as a result of warming seas. Increasing acidity in the oceans is another long-term threat to marine ecosystems. Ice-based ecologies have also suffered devastating climate change impacts, especially in the Arctic region. While some animal and plant species will adapt, for many species the pace of climate change is too rapid: climate systems are moving more rapidly than they can follow. With 3°C of warming, 20–30 percent of land species could face extinction.

- Human health. Rich countries are already preparing public health systems to deal with future climate shocks, such as the 2003 European heatwave and more extreme summer and winter conditions. However, the greatest health impacts will be felt in developing countries because of high levels of poverty and the limited capacity of public health systems to respond. Major killer diseases could expand their coverage. For example, an additional 220–400 million people could be exposed to malaria—a disease that already claims around 1 million lives annually. Dengue fever is already in evidence at higher levels of elevation than has previously been the case, especially parts of East Asia. Climate change could further expand the reach of the disease.

None of these five separate drivers will operate in isolation. They will interact with wider social, economic and ecological processes that shape opportunities for human development.

Inevitably, the precise mix of transmission mechanisms from climate change to human development will vary across and within countries. Large areas of uncertainty remain. What is certain is that dangerous climate change has the potential to deliver powerful systemic shocks to human development across a large group of countries. In contrast to economic shocks that affect growth or inflation, many of the human development impacts—lost opportunities for health and education, diminished productive potential, loss of vital ecological systems, for example—are likely to prove irreversible.

Avoiding dangerous climate change: strategies for mitigation

Avoiding the unprecedented threats posed by dangerous climate change will require an unparalleled collective exercise in international cooperation. Negotiations on emission limits for the post-2012 Kyoto Protocol commitment period can—and must—frame the global carbon budget. However, a sustainable global emissions pathway will only be meaningful if it is translated into practical national strategies—and national carbon budgets. Climate change mitigation is about transforming the way that we produce and use energy. And it is about living within the bounds of ecological sustainability.

Setting credible targets linked to global mitigation goals is the starting point for the transition to a sustainable emissions pathway. These targets can provide a basis for carbon budgeting exercises that provide a link from the present to the future through a series of rolling plans. However, credible targets have to be backed by clear policies. The record to date in this area is not encouraging. Most developed countries are falling short of the targets set under the Kyoto Protocol: Canada is an extreme case in point. In some cases, ambitious 'Kyoto-plus' targets have been adopted. The European Union and the United Kingdom have both embraced such targets. For different reasons, they are both likely to fall far short of the goals set unless they move rapidly to put climate mitigation at the centre of energy policy reform.

Two major OECD countries are not bound by Kyoto targets. Australia has opted for a wide-ranging voluntary initiative, which has produced mixed results. The United States does not have a federal target for reducing emissions. Instead, it has a 'carbon-intensity' reduction goal which measures efficiency. The problem is that efficiency gains have failed to prevent large aggregate increases in emissions. In the absence of federal targets, several United States' states have set their own mitigation goals. California's Global Warming Solutions Act of 2006 is a bold attempt to align greenhouse gas reduction targets with reformed energy policies.

Setting ambitious targets for mitigation is an important first step. Translating targets into policies is politically more challenging. The starting point: putting a price on carbon emissions. Changed incentive structures are a vital condition for an accelerated transition to low-carbon growth. In an optimal scenario, the carbon price would be global. This is politically unrealistic in the short-run because the world lacks the required governance system. The more realistic option is for rich countries to develop carbon pricing structures. As these structures evolve, developing countries could be integrated over time as institutional conditions allow.

Enhanced energy efficiency has the potential to deliver a 'double dividend'. It can reduce CO_2 emissions and cut energy costs. If all electrical appliances operating in OECD countries in 2005 had met the best efficiency standards, it would have saved some 322 Mt CO_2 of emissions by 2010—equivalent to taking over 100 million cars off the road. Household electricity consumption would fall by one-quarter. Personal transportation is another area where regulatory standards can unlock double-dividends. The automobile sector accounts for about 30 percent of greenhouse gas emissions in developed countries—and the share is rising. Regulatory standards matter because they can influence fleet efficiency, or the average number of miles travelled per gallon (and hence CO_2 emissions). In the United States, fuel efficiency standards have slipped over time. They are now lower than in China. Efforts to raise fuel efficiency standards are often countered by powerful vested interests. In Europe, for example, European Commission proposals to raise standards have been countered by a coalition of automobile manufacturers. Several member states have rejected the proposals, raising wider questions about the European Union's capacity to translate climate change goals into tangible policies.

International trade could play a much larger role in expanding markets for alternative fuels. Brazil is more efficient than either the European Union or the United States in producing ethanol. Moreover, sugar-based ethanol is more efficient at cutting carbon emissions. The problem is that imports of Brazilian ethanol are restricted by high import tariffs. Removing these tariff s would generate gains not just for Brazil, but for climate change mitigation.

The rapid development and deployment of low-carbon technologies is vital to climate change mitigation. Picking winners in technology is a hazardous affair. Governments have at best a mixed record. However, confronted with a national and global threat on the scale of climate change, governments cannot afford to stand back and wait for markets to deliver. Energy policy is an area in which the scale of upfront investments, time horizon, and uncertainty combine to guarantee that markets alone will fail to deliver technological change at the pace required by mitigation. In earlier periods, major technological breakthroughs have followed decisive government action: the Manhattan Project and the United States space programme are examples.

Carbon Capture and Storage (CSS) is a key breakthrough technology. Coal is the major source of power for electricity generation worldwide. Reserves are widely dispersed. Coupled with rising prices for oil and natural gas, this is one reason why coal figures prominently in the present and planned energy mix of major emitters such as the China, India and the United States. CCS is important because it holds out the promise of coal-fired power generation with near-zero emissions. With a more active programme of public–private investment, aligned with carbon pricing, CCS technologies could be developed and deployed more rapidly.

Low levels of energy efficiency in developing countries are currently a threat to climate change mitigation efforts. Raising efficiency levels through international cooperation could transform that threat into an opportunity, generating large gains for human development in the process. We demonstrate this by examining the impact on CO_2 emissions of an accelerated technology transfer programme for the coal sector in China. For China alone, emissions in 2030 would be 1.8 Gt CO_2 below the level projected by the International Energy Agency. That figure is equivalent to around one-half of current European Union emissions. Similar efficiency gains are attainable in other areas. Enhanced energy efficiency is a win–win scenario. Developing countries stand to gain from improved energy efficiency and lower environmental pollution. All countries stand to gain from CO_2 mitigation. Unfortunately, the world currently lacks a credible mechanism for unlocking this win–win scenario. We propose the development, under the auspices of the post-2012 Kyoto framework, of a Climate Change Mitigation Facility (CCMF) to fill this gap. The CCMF would mobilize US$25–50 billion annually to finance low-carbon energy investments in developing countries. Financing provisions would be linked to the circumstances of individual countries, with a menu of grants, concessional support and risk guarantees available. Support would be programme based. It would cover the incremental costs of achieving defined emission reduction targets by scaling-up nationally-owned energy policies in areas such as renewable energy, clean coal and enhanced efficiency standards for transport and buildings.

Deforestation is another key area for international cooperation. Currently, the world is losing the carbon assets contained in rainforests at a fraction of the market value they would have even at low carbon prices. In Indonesia, every US$1 generated through deforestation to grow palm oil would translate into a US$50–100 loss if the reduced carbon capacity could be traded on the European Union's ETS. Beyond these market failures, the loss of rainforests represents the erosion of a resource that plays a vital role in the lives of the poor, in the provision of ecosystem services and in sustaining biodiversity.

There is scope for exploring the potential of carbon markets in the creation of incentives to avoid deforestation. More broadly, carbon finance could be mobilized to support the restoration of degraded grasslands, generating benefits for climate change mitigation, adaptation and environmental sustainability.

Adapting to the inevitable: national action and international cooperation

Without urgent mitigation action the world cannot avoid dangerous climate change. But even the most stringent mitigation will be insufficient to avoid major human development setbacks. The world is already committed to further warming because of the inertia built into climate systems and the delay between mitigation and outcome. For the first half of the 21st Century there is no alternative to adaptation to climate change.

Rich countries already recognize the imperative to adapt. Many are investing heavily in the development of climate defence infrastructures. National strategies are being drawn up to prepare for more extreme and less certain future weather patterns.

Developing countries face far more severe adaptation challenges. Those challenges have to be met by governments operating under severe financing constraints, and by poor people themselves. In the Horn of Africa, 'adaptation' means that women and young girls walk further to collect water. In the Ganges Delta, people are erecting bamboo flood shelters on stilts. And in the Mekong Delta people are planting mangroves to protect themselves against storm surges, and women and children are being taught to swim.

Inequalities in capacity to adapt to climate change are becoming increasingly apparent. For one part of the world—the richer part—adaptation is a matter of erecting elaborate climate defence infrastructures, and of building homes that 'float on' water. In the other part adaptation means people themselves learning to 'float in' flood water. Unlike people living behind the flood defences of London and Los Angeles, young girls in the Horn of Africa and people in the Ganges Delta do not have a deep carbon footprint. As Desmond Tutu, the former Archbishop of Cape Town, has argued, we are drifting into a world of adaptation apartheid.

Planning for climate change adaptation confronts governments in developing countries with challenges at many levels. These challenges pose systemic threats. In Egypt, delta flooding could transform conditions for agricultural production. Changes to coastal currents in southern Africa could compromise the future of Namibia's fisheries sector. Hydroelectric power generation will be affected in many countries. Responding to climate change will require the integration of adaptation into all aspects of policy development and planning for poverty reduction. However, planning and implementation capacity is limited:

- Information. Many of the world's poorest countries lack the capacity and the resources to assess climate risks. In sub-Saharan Africa, high levels of rural poverty and dependence on rainfed agriculture makes meteorological information an imperative for adaptation. However, the region has the world's lowest density of meteorological stations. In France, the meteorological budget amounts to US$388 million annually, compared with just US$2 million in Ethiopia. The 2005 G8 summit pledged action to strengthen Africa's meteorological monitoring capacity.

  Follow-up has fallen far short of the commitments made.

- Infrastructure. In climate change adaptation, as in other areas, prevention is better than cure. Every US$1 invested in pre-disaster risk management in developing countries can prevent losses of US$7. In Bangladesh, research among impoverished populations living on char islands shows that adaptation against flooding can strengthen livelihoods, even in extreme conditions. Many countries lack the financial resources required for infrastructural adaptation. Beyond disaster prevention, the development of community based infrastructure for water harvesting can reduce vulnerability and empower people to cope with climate risks. Partnerships between communities and local governments in Indian states such as Andhra Pradesh and Gujarat provide examples of what can be achieved.

- Insurance for social protection. Climate change is generating incremental risks in the lives of the poor. Social protection programmes can help people cope with those risks while expanding opportunities for employment, nutrition and education. In Ethiopia the Productive Safety Net Programme is an attempt to strengthen the capacity of poor households to cope with droughts without having to sacrifice opportunities for health and education. In Latin America conditional cash transfers have been widely used to support a wide range of human development goals, including the protection of basic capabilities during a sudden crisis. In southern Africa cash transfers have been used during droughts to protect long-run productive capacity. While social protection figures only marginally in current climate change adaptation strategies, it has the potential to create large human development returns.

The case for international action on adaptation is rooted in past commitments, shared values, the global commitment to poverty reduction and the liability of rich nations for climate change problems. Under the terms of the United Nations Framework Convention on Climate Change (UNFCCC), northern governments are obliged to support adaptation capacity development. Support for the MDGs provides another powerful rationale for action: adaptation is a key requirement for achieving the 2015 targets and creating the conditions for sustained progress. Application of the legal principles of protection from harm and compensation for damage would constitute further grounds for action.

Expressed in diplomatic language, the international response on adaptation has fallen far short of what is required. Several dedicated multilateral financing mechanisms have been created, including the Least Developed Country Fund and the Special Climate Change Fund. Delivery through these mechanisms has been limited. Total financing to date has amounted to around US$26 million—a derisory response. For purposes of comparison, this is equivalent to one week's worth of spending under the United Kingdom flood defence programme. Current pledged funding amounts to US$279 million for disbursement over several years. This is an improvement over past delivery but still a fraction of what is required. It represents less than one-half of what the German state of Baden-Würtemberg will allocate to the strengthening of flood defences.

It is not just the lives and the livelihoods of the poor that require protection through adaptation. Aid programmes are also under threat. We estimate that around one-third of current development assistance is concentrated in areas facing varying degrees of climate change risk. Insulating aid budgets from that risk will require additional investment of around US$4.5 billion. At the same time, climate change is contributing to a diversion of aid into disaster relief. This has been one of the fastest growing areas for aid flows, accounting for 7.5 percent of total commitments in 2005.

Estimating the aid financing requirements for adaptation is inherently difficult. In the absence of detailed national assessments of climate change risks and vulnerabilities, any assessment must remain a 'guesstimate'. Our 'guesstimate' is that by 2015 at least US$44 billion will be required annually for 'climate proofing' development investments (2005 prices). Building human resilience is another priority area. Investments in social protection and wider human development strategies are needed to strengthen the capacity of vulnerable people to cope with risk. Our ballpark estimate is that at least US$40 billion will be needed by 2015 to strengthen national strategies for poverty reduction in the face of climate change risks. To put this figure in context, it represents around 0.5 percent of projected 2015 GDP for low income and lower middle income countries. Provision for disaster and post-disaster recovery will also have to be strengthened as droughts, floods, storms and landslides pose greater threats. Provision of an additional US$2 billion a year is implied by our estimates.

Adaptation financing requirements should be seen as 'new and additional' commitments. That is, they should supplement rather than divert existing aid commitments. Northern governments have pledged to double aid by 2010, though the record on delivery is mixed. Any shortfall in delivery will compromise progress towards the MDGs and compound problems in climate change adaptation.

The headline figure for new and additional adaptation financing appears large—but has to be placed in context. The total of around US$86 billion by 2015 may be required to prevent aid diversion. It would represent around 0.2 percent of developed country GDP, or around one-tenth of what they currently allocate to military expenditure. Measured in terms of returns for human security, adaptation financing is a highly cost-effective investment. There are a range of innovative financing mechanisms that could be explored to mobilize resources. These include carbon taxation, levies administered under cap-and-trade programmes and dedicated levies on air transport and vehicles.

International support for adaptation has to go beyond financing. Current international efforts suffer not just from chronic underfinancing, but also a lack of coordination and coherence. The patchwork of multilateral mechanisms is delivering small amounts of finance with very high transaction costs, most of it through individual projects. While project-based support has an important role to play, the locus for adaptation planning has to be shifted towards national programmes and budgets.

The integration of adaptation planning into wider poverty reduction strategies is a priority. Successful adaptation policies cannot be grafted on to systems that are failing to address underlying causes of poverty, vulnerability and wider disparities based on wealth, gender and location.

Conclusion and summary of recommendations

Climate change confronts humanity with stark choices. We can avoid 21st Century reversals in human development and catastrophic risks for future generations, but only by choosing to act with a sense of urgency. That sense of urgency is currently missing. Governments may use the rhetoric of a 'global security crisis' when describing the climate change problem, but their actions—and inactions—on energy policy reform tell a different story. The starting point for action and political leadership is recognition on the part of governments that they are confronted by what may be the gravest threat ever to have faced humanity.

Facing up to that threat will create challenges at many levels. Perhaps most fundamentally of all, it challenges the way that we think about progress. There could be no clearer demonstration than climate that economic wealth creation is not the same thing as human progress. Under the current energy policies, rising economic prosperity will go hand-in-hand with mounting threats to human development today and the well-being of future generations. But carbon-intensive economic growth is symptomatic of a deeper problem. One of the hardest lessons taught by climate change is that the economic model which drives growth, and the profligate consumption in rich nations that goes with it, is ecologically unsustainable. There could be no greater challenge to our assumptions about progress than that of realigning economic activities and consumption with ecological realities.

Combating climate change demands that we place ecological imperatives at the heart of economics. That process has to start in the developed world—and it has to start today. The uncertainties have to be acknowledged. In this report we have argued that, with the right reforms, it is not too late to cut greenhouse gas emissions to sustainable levels without sacrificing economic growth: that rising prosperity and climate security are not conflicting objectives.

The current state of international cooperation and multilateralism on climate change is not fit for the purpose. As a priority, the world needs a binding international agreement to cut greenhouse gas emissions across a long time horizon, but with stringent near-term and medium-term targets. The major developing countries have to be party to that agreement and make commitments to reduce emissions. However, those commitments will need to reflect their circumstances and capabilities, and the overarching need to sustain progress in poverty reduction. Any multilateral agreement without quantitative commitments from developing countries will lack credibility in terms of climate change mitigation. At the same time, no such agreement will emerge unless it incorporates provisions for finance and technology transfer from the rich nations that bear historic responsibility for climate change.

International cooperation must also address the pressing issue of climate change adaptation. Even with stringent mitigation, the world is already committed to sustained global warming for the first half of the 21st Century. Having created the problem, the world's richest countries cannot stand aside and watch the hopes and the aspirations of the world's poor be undermined by increased exposure to the risks and vulnerabilities that will come with climate change.

Fighting climate change is a cross-generational exercise. For the current generation, the challenge is to keep open the window of opportunity by bending greenhouse gas emissions in a downward direction. The world has a historic opportunity to begin this task. In 2012, the current commitment period of the Kyoto Protocol expires. The successor agreement could set a new course, imposing stringent limits on future emissions and providing a framework for international collective action. Negotiations could be brought forward so that the quantitative targets are set by 2010, providing governments with goals for national carbon budgets. Carbon budgeting backed by radical energy policy reforms and government action to change incentive structures for consumers and investors is the foundation for effective climate change mitigation. There is no such thing as a last chance in human affairs. But the post-2012 Kyoto framework comes close.

Recommendations

1 **Develop a multilateral framework for avoiding dangerous climate change under the post-2012 Kyoto Protocol**

- Establish an agreed threshold for dangerous climate change at 2°C above preindustrial levels.

- Set a stabilization target for atmospheric concentrations of CO_2e at 450 ppm (the costs are estimated at 1.6 percent of average global GDP to 2030).

- Agree to a global sustainable emissions pathway aimed at 50 percent reductions of greenhouse gas emissions by 2050 from 1990 levels.

- Targets under the current Kyoto commitment period implemented by developed countries, with a further agreement to cut greenhouse gas emissions by at least 80 percent by 2050, with 20–30 percent cuts by 2020.

- Major emitters in developing countries to aim at an emissions trajectory that peaks in 2020, with 20 percent cuts by 2050.

2 **Put in place policies for sustainable carbon budgeting— the agenda for mitigation**

- Set a national carbon budget in all developed countries with targets for reducing overall emissions from a 1990 reference year incorporated into national legislation.

- Put a price on carbon through taxation or cap-and-trade programmes consistent with national carbon budget goals.

- Carbon taxation to be introduced at a level of US$10–20/t CO_2 in 2010, rising in annual increments to US$60–100/t CO_2.

- Adopt cap-and-trade programmes that aim at 20–30 percent cuts in CO_2 emissions by 2020 with 90–100 percent of allowances auctioned by 2015.

- Utilise revenues from carbon taxation and cap-and-trade to fi nance progressive tax reform, with reductions in taxes on labour and investments, and the development of incentives for low-carbon technology.

- Reform of the European Union's Emissions Trading Scheme to reduce quotas, increase auctioning and limit windfall gains for the private sector.

- Create an enabling environment for renewable energy through 'feed-in' tariff s and market regulation, with a 20 percent target by 2020 in renewable power generation.

- Increase energy efficiency through regulatory standards for appliances and buildings.

- Reduce CO_2 emissions from transport through stronger fuel efficiency standards in the European Union, with a target of 120g CO_2/km by 2012 and 80g CO_2/km by 2020, and more stringent Corporate Average Fuel Economy Standards (CAFE) in the United States with the introduction of taxation of aviation.

- Increase financing, incentives and regulatory support for the development of breakthrough technologies, with a focus on Carbon Capture and Storage (CCS)—the United States should aim at 30 demonstration plants by 2015, and the European Union should have a comparable level of ambition.

3. Strengthen the framework for international cooperation

- Develop international cooperation to enhance access to modern energy services and reduce dependence on biomass, the primary source of energy for about 2.5 billion people.

- Reduce the rate of increase in carbon emissions in developing countries through strengthened energy sector reforms, backed by finance and technology transfer.

- Create a Climate Change Mitigation Facility (CCMF) to mobilize the US$25–50 billion needed annually to support low-carbon transitions in developing countries through a mix of grants, concessional aid and risk guarantees for investment under nationally-owned energy sector reform programmes.

- Integrate project based carbon-financing through the Clean Development Mechanism and other Kyoto flexibility provisions into programme-based and sectoral national strategies for supporting low-carbon transition.

- Significantly strengthen international cooperation on coal, with the creation of incentives for the development and deployment on Integrated Gasification Combined Cycle (IGCC) technology and CCS.

- Develop international incentives for the conservation and sustainable management of rainforests.

- Extend carbon financing beyond industrial sector mitigation to land-use programmes—such as forest conservation and grasslands restoration—that offer benefits for the poor.

4 Put climate change adaptation at the centre of the post-2012 Kyoto framework and international partnerships for poverty reduction

- Recognize that the world is committed to significant climate change, that even stringent mitigation will not materially affect temperature change until the mid-2030s, and that average global temperatures will rise to 2050 even under a 'good case' scenario.

- Strengthen the capacity of developing countries to assess climate change risks and integrate adaptation into all aspects of national planning.

- Act on G8 commitments to strengthen meteorological monitoring capacity in sub- Saharan Africa through partnerships under the Global Climate Observing System.

- Empower and enable vulnerable people to adapt to climate change by building resilience through investments in social protection, health, education and other measures.
- Integrate adaptation into poverty reduction strategies that address vulnerabilities linked to inequalities based on wealth, gender, location and other markers for disadvantage.
- Provide at least US$86 billion in 'new and additional' finance for adaptation through transfers from rich to poor by 2016 to protect progress towards the MDGs and prevent post-2015 reversals in human development.
- Expand multilateral provisions for responding to climate-related humanitarian emergencies and supporting post-disaster recovery to build future resilience, with US$2 billion in financing by 2016 under arrangements such as the UN's Central Emergency Response Fund and the World Bank's Global Facility for Disaster Reduction and Recovery.
- Explore a range of innovative financing options beyond development assistance to mobilize support for adaptation, including carbon taxation, levies on quotas issued under cap-and-trade programmes, air transport taxes and wider measures.
- Streamline the current structure of dedicated multilateral funds which are delivering limited support (US$26 million to date and US$253 million in the pipeline, with high transition costs), and shift the locus of support from projects to programme-based financing.
- Use Poverty Reduction Strategy Papers (PRSPs) to conduct national estimates of the costs of scaling-up existing programmes, identifying priority areas for reducing vulnerability.

Appendix -III

United Nations Framework Convention on Climate Change

Table of Contents

- Article 17 Protocols
- Article 18 Right to Vote
- Article 19 Depositary
- Article 20 Signature
- Article 21 Interim Arrangements
- Article 22 Ratification, Acceptance, Approval or Accession
- Article 23 Entry into Force
- Article 24 Reservations
- Article 25 Withdrawal

Preamble

The Parties to this Convention,

Acknowledging that change in the Earth's climate and its adverse effects are a common concern of humankind,

Concerned that human activities have been substantially increasing the atmospheric concentrations of greenhouse gases, that these increases enhance the natural greenhouse effect, and that this will result on average in an additional warming of the Earth's surface and atmosphere and may adversely affect natural ecosystems and humankind,

Noting that the largest share of historical and current global emissions of greenhouse gases has originated in developed countries, that per capita emissions in developing countries are still relatively low and that the share of global emissions originating in developing countries will grow to meet their social and development needs,

Aware of the role and importance in terrestrial and marine ecosystems of sinks and reservoirs of greenhouse gases,

Noting that there are many uncertainties in predictions of climate change, particularly with regard to the timing, magnitude and regional patterns thereof,

Acknowledging that the global nature of climate change calls for the widest possible cooperation by all countries and their participation in an effective and appropriate international response, in accordance with their common but differentiated responsibilities and respective capabilities and their social and economic conditions,

Recalling the pertinent provisions of the Declaration of the United Nations Conference on the Human Environment, adopted at Stockholm on 16 June 1972,

Recalling also that States have, in accordance with the Charter of the United Nations and the principles of international law, the sovereign right to exploit their own resources pursuant to their own environmental and developmental policies, and the responsibility to ensure that activities within their jurisdiction or control do not cause damage to the environment of other States or of areas beyond the limits of national jurisdiction,

Reaffirming the principle of sovereignty of States in international cooperation to address climate change,

Recognizing that States should enact effective environmental legislation, that environmental standards, management objectives and priorities should reflect the environmental and developmental context to which they apply, and that standards applied by some countries may be inappropriate and of unwarranted economic and social cost to other countries, in particular developing countries,

Recalling the provisions of General Assembly resolution 44/228 of 22 December 1989 on the United Nations Conference on Environment and Development, and resolutions 43/53 of 6 December 1988, 44/207 of 22 December 1989, 45/212 of 21 December 1990 and 46/169 of 19 December 1991 on protection of global climate for present and future generations of mankind,

Recalling also the provisions of General Assembly resolution 44/206 of 22 December 1989 on the possible adverse effects of sea-level rise on islands and coastal areas, particularly low-lying coastal areas and the pertinent provisions of General Assembly resolution 44/172 of 19 December 1989 on the implementation of the Plan of Action to Combat Desertification,

Recalling further the Vienna Convention for the Protection of the Ozone Layer, 1985, and the Montreal Protocol on Substances that Deplete the Ozone Layer, 1987, as adjusted and amended on 29 June 1990,

Noting the Ministerial Declaration of the Second World Climate Conference adopted on 7 November 1990,

Conscious of the valuable analytical work being conducted by many States on climate change and of the important contributions of the World Meteorological Organization, the United Nations Environment Programme and other organs, organizations and bodies of the United Nations system, as well as other international and intergovernmental bodies, to the exchange of results of scientific research and the coordination of research,

Recognizing that steps required to understand and address climate change will be environmentally, socially and economically most effective if they are based on relevant scientific, technical and economic considerations and continually re-evaluated in the light of new findings in these areas,

Recognizing that various actions to address climate change can be justified economically in their own right and can also help in solving other environmental problems,

Recognizing also the need for developed countries to take immediate action in a flexible manner on the basis of clear priorities, as a first step towards comprehensive response strategies at the global, national and, where agreed, regional levels that take into account all greenhouse gases, with due consideration of their relative contributions to the enhancement of the greenhouse effect,

Recognizing further that low-lying and other small island countries, countries with low-lying coastal, arid and semi-arid areas or areas liable to floods, drought and desertification, and developing countries with fragile mountainous ecosystems are particularly vulnerable to the adverse effects of climate change,

Recognizing the special difficulties of those countries, especially developing countries, whose economies are particularly dependent on fossil fuel production, use and exportation, as a consequence of action taken on limiting greenhouse gas emissions,

Affirming that responses to climate change should be coordinated with social and economic development in an integrated manner with a view to avoiding adverse impacts on the latter, taking into full account the legitimate priority needs of developing countries for the achievement of sustained economic growth and the eradication of poverty,

Recognizing that all countries, especially developing countries, need access to resources required to achieve sustainable social and economic development and that, in order for developing countries to progress towards that goal, their energy consumption will need to grow taking into account the possibilities for achieving greater energy efficiency and for controlling greenhouse gas emissions in general, including through the application of new technologies on terms which make such an application economically and socially beneficial,

Determined to protect the climate system for present and future generations,

Have agreed as follows:

Article 1: Definitions

For the purposes of this Convention:

1. "Adverse effects of climate change" means changes in the physical environment or biota resulting from climate change which have significant deleterious effects on the composition, resilience or productivity of natural and managed ecosystems or on the operation of socio-economic systems or on human health and welfare.

2. "Climate change" means a change of climate which is attributed directly or indirectly to human activity that alters the composition of the global atmosphere and which is in addition to natural climate variability observed over comparable time periods.

3. "Climate system" means the totality of the atmosphere, hydrosphere, biosphere and geosphere and their interactions.

4. "Emissions" means the release of greenhouse gases and/or their precursors into the atmosphere over a specified area and period of time.

5. "Greenhouse gases" means those gaseous constituents of the atmosphere, both natural and anthropogenic, that absorb and re-emit infrared radiation.

6. "Regional economic integration organization" means an organization constituted by sovereign States of a given region which has competence in respect of matters governed by this Convention or its protocols and has been duly authorized, in accordance with its internal procedures, to sign, ratify, accept, approve or accede to the instruments concerned.

7. "Reservoir" means a component or components of the climate system where a greenhouse gas or a precursor of a greenhouse gas is stored.

8. "Sink" means any process, activity or mechanism which removes a greenhouse gas, an aerosol or a precursor of a greenhouse gas from the atmosphere.

9. "Source" means any process or activity which releases a greenhouse gas, an aerosol or a precursor of a greenhouse gas into the atmosphere.

Article 2: Objective

The ultimate objective of this Convention and any related legal instruments that the Conference of the Parties may adopt is to achieve, in accordance with the relevant provisions of the Convention, stabilization of greenhouse gas concentrations in the atmosphere at a level that would prevent dangerous anthropogenic interference with the climate system. Such a level should be achieved within a time-frame sufficient to allow ecosystems to adapt naturally to climate change, to ensure that food production is not threatened and to enable economic development to proceed in a sustainable manner.

Article 3: Principles

In their actions to achieve the objective of the Convention and to implement its provisions, the Parties shall be guided, INTER ALIA, by the following:

1. The Parties should protect the climate system for the benefit of present and future generations of humankind, on the basis of equity and in accordance with their common but differentiated responsibilities and respective capabilities. Accordingly, the developed country Parties should take the lead in combating climate change and the adverse effects thereof.

2. The specific needs and special circumstances of developing country Parties, especially those that are particularly vulnerable to the adverse effects of climate change, and of those Parties, especially developing country Parties, that would have to bear a disproportionate or abnormal burden under the Convention, should be given full consideration.

3. The Parties should take precautionary measures to anticipate, prevent or minimize the causes of climate change and mitigate its adverse effects. Where there are threats of serious or irreversible damage, lack of full scientific certainty should not be used as a reason for postponing such measures, taking into account that policies and measures to deal with climate change should be cost-effective so as to ensure global benefits at the lowest possible cost. To achieve this, such policies and measures should take into account different socio-economic contexts, be comprehensive, cover all relevant sources, sinks and reservoirs of greenhouse gases and adaptation, and comprise all economic sectors. Efforts to address climate change may be carried out cooperatively by interested Parties.

4. The Parties have a right to, and should, promote sustainable development. Policies and measures to protect the climate system against human-induced change should be appropriate for the specific conditions of each Party and should be integrated with national development programmes, taking into account that economic development is essential for adopting measures to address climate change.

5. The Parties should cooperate to promote a supportive and open international economic system that would lead to sustainable economic growth and development in all

Parties, particularly developing country Parties, thus enabling them better to address the problems of climate change. Measures taken to combat climate change, including unilateral ones, should not constitute a means of arbitrary or unjustifiable discrimination or a disguised restriction on international trade.

Article 4: Commitments

1. All Parties, taking into account their common but differentiated responsibilities and their specific national and regional development priorities, objectives and circumstances, shall:

2. (a) Develop, periodically update, publish and make available to the Conference of the Parties, in accordance with Article 12, national inventories of anthropogenic emissions by sources and removals by sinks of all greenhouse gases not controlled by the Montreal Protocol, using comparable methodologies to be agreed upon by the Conference of the Parties;

(b) Formulate, implement, publish and regularly update national and, where appropriate, regional programmes containing measures to mitigate climate change by addressing anthropogenic emissions by sources and removals by sinks of all greenhouse gases not controlled by the Montreal Protocol, and measures to facilitate adequate adaptation to climate change;

(c) Promote and cooperate in the development, application and diffusion, including transfer, of technologies, practices and processes that control, reduce or prevent anthropogenic emissions of greenhouse gases not controlled by the Montreal Protocol in all relevant sectors, including the energy, transport, industry, agriculture, forestry and waste management sectors;

(d) Promote sustainable management, and promote and cooperate in the conservation and enhancement, as appropriate, of sinks and reservoirs of all greenhouse gases not controlled by the Montreal Protocol, including biomass, forests and oceans as well as other terrestrial, coastal and marine ecosystems;

(e) Cooperate in preparing for adaptation to the impacts of climate change; develop and elaborate appropriate and integrated plans for coastal zone management, water resources and agriculture, and for the protection and rehabilitation of areas, particularly in Africa, affected by drought and desertification, as well as floods;

(f) Take climate change considerations into account, to the extent feasible, in their relevant social, economic and environmental policies and actions, and employ appropriate methods, for example impact assessments, formulated and determined nationally, with a view to minimizing adverse effects on the economy, on public health and on the quality of the environment, of projects or measures undertaken by them to mitigate or adapt to climate change;

(g) Promote and cooperate in scientific, technological, technical, socio-economic and other research, systematic observation and development of data archives related to the climate system and intended to further the understanding and to reduce or eliminate the remaining uncertainties regarding the causes, effects, magnitude and timing of climate change and the economic and social consequences of various response strategies;

(h) Promote and cooperate in the full, open and prompt exchange of relevant scientific, technological, technical, socio-economic and legal information related to the climate system and climate change, and to the economic and social consequences of various response strategies;

(i) Promote and cooperate in education, training and public awareness related to climate change and encourage the widest participation in this process, including that of non-governmental organizations; and

(j) Communicate to the Conference of the Parties information related to implementation, in accordance with Article 12.

3. The developed country Parties and other Parties included in Annex I commit themselves specifically as provided for in the following:

(a) Each of these Parties shall adopt national policies and take corresponding measures on the mitigation of climate change, by limiting its anthropogenic emissions of greenhouse gases and protecting and enhancing its greenhouse gas sinks and reservoirs. These policies and measures will demonstrate that developed countries are taking the lead in modifying longer-term trends in anthropogenic emissions consistent with the objective of the Convention, recognizing that the return by the end of the present decade to earlier levels of anthropogenic emissions of carbon dioxide and other greenhouse gases not controlled by the Montreal Protocol would contribute to such modification, and taking into account the differences in these Parties' starting points and approaches, economic structures and resource bases, the need to maintain strong and sustainable economic growth, available technologies and other individual circumstances, as well as the need for equitable and appropriate contributions by each of these Parties to the global effort regarding that objective. These Parties may implement such policies and measures jointly with other Parties and may assist other Parties in contributing to the achievement of the objective of the Convention and, in particular, that of this subparagraph;

(b) In order to promote progress to this end, each of these Parties shall communicate, within six months of the entry into force of the Convention for it and periodically thereafter, and in accordance with Article 12, detailed information on its policies and measures referred to in subparagraph (a) above, as well as on its resulting projected anthropogenic emissions by sources and removals by sinks of greenhouse gases not controlled by the Montreal Protocol for the period referred to in subparagraph (a), with the aim of returning individually or jointly to their 1990 levels these anthropogenic emissions of carbon dioxide and other greenhouse gases not controlled by the Montreal Protocol. This information will be reviewed by the Conference of the Parties, at its first session and periodically thereafter, in accordance with Article 7;

(c) Calculations of emissions by sources and removals by sinks of greenhouse gases for the purposes of subparagraph (b) above should take into account the best available scientific knowledge, including of the effective capacity of sinks and the respective contributions of such gases to climate change. The Conference of the Parties shall consider and agree on methodologies for these calculations at its first session and review them regularly thereafter;

(d) The Conference of the Parties shall, at its first session, review the adequacy of subparagraphs (a) and (b) above. Such review shall be carried out in the light of the best available scientific information and assessment on climate change and its impacts, as well as relevant technical, social and economic information. Based on this review, the Conference of the Parties shall take appropriate action, which may include the adoption of amendments to the commitments in subparagraphs (a) and (b) above. The Conference of the Parties, at its first session, shall also take decisions regarding criteria for joint implementation as indicated in subparagraph (a) above. A second review of subparagraphs (a) and (b) shall take place not later than 31 December 1998, and thereafter at regular intervals determined by the Conference of the Parties, until the objective of the Convention is met;

(e) Each of these Parties shall:

(i) Coordinate as appropriate with other such Parties, relevant economic and administrative instruments developed to achieve the objective of the Convention; and

(ii) Identify and periodically review its own policies and practices which encourage activities that lead to greater levels of anthropogenic emissions of greenhouse gases not controlled by the Montreal Protocol than would otherwise occur;

(f) The Conference of the Parties shall review, not later than 31 December 1998, available information with a view to taking decisions regarding such amendments to the lists in Annexes I and II as may be appropriate, with the approval of the Party concerned;

(g) Any Party not included in Annex I may, in its instrument of ratification, acceptance, approval or accession, or at any time thereafter, notify the Depositary that it intends to be bound by subparagraphs (a) and (b) above. The Depositary shall inform the other signatories and Parties of any such notification.

4. The developed country Parties and other developed Parties included in Annex II shall provide new and additional financial resources to meet the agreed full costs incurred by developing country Parties in complying with their obligations under Article 12, paragraph 1. They shall also provide such financial resources, including for the transfer of technology, needed by the developing country Parties to meet the agreed full incremental costs of implementing measures that are covered by paragraph 1 of this Article and that are agreed between a developing country Party and the international entity or entities referred to in Article 11, in accordance with that Article. The

implementation of these commitments shall take into account the need for adequacy and predictability in the flow of funds and the importance of appropriate burden sharing among the developed country Parties.

5. The developed country Parties and other developed Parties included in Annex II shall also assist the developing country Parties that are particularly vulnerable to the adverse effects of climate change in meeting costs of adaptation to those adverse effects.

6. The developed country Parties and other developed Parties included in Annex II shall take all practicable steps to promote, facilitate and finance, as appropriate, the transfer of, or access to, environmentally sound technologies and know-how to other Parties, particularly developing country Parties, to enable them to implement the provisions of the Convention. In this process, the developed country Parties shall support the development and enhancement of endogenous capacities and technologies of developing country Parties. Other Parties and organizations in a position to do so may also assist in facilitating the transfer of such technologies.

7. In the implementation of their commitments under paragraph 2 above, a certain degree of flexibility shall be allowed by the Conference of the Parties to the Parties included in Annex I undergoing the process of transition to a market economy, in order to enhance the ability of these Parties to address climate change, including with regard to the historical level of anthropogenic emissions of greenhouse gases not controlled by the Montreal Protocol chosen as a reference.

8. The extent to which developing country Parties will effectively implement their commitments under the

Convention will depend on the effective implementation by developed country Parties of their commitments under the Convention related to financial resources and transfer of technology and will take fully into account that economic and social development and poverty eradication are the first and overriding priorities of the developing country Parties.

9. In the implementation of the commitments in this Article, the Parties shall give full consideration to what actions are necessary under the Convention, including actions related to funding, insurance and the transfer of technology, to meet the specific needs and concerns of developing country Parties arising from the adverse effects of climate change and/or the impact of the implementation of response measures, especially on:

(a) Small island countries;

(b) Countries with low-lying coastal areas;

(c) Countries with arid and semi-arid areas, forested areas and areas liable to forest decay;

(d) Countries with areas prone to natural disasters;

(e) Countries with areas liable to drought and desertification;

(f) Countries with areas of high urban atmospheric pollution;

(g) Countries with areas with fragile ecosystems, including mountainous ecosystems;

(h) Countries whose economies are highly dependent on income generated from the production, processing and export, and/or on consumption of fossil fuels and associated energy-intensive products; and

(i) Land-locked and transit countries.

Further, the Conference of the Parties may take actions, as appropriate, with respect to this paragraph.

10. The Parties shall take full account of the specific needs and special situations of the least developed countries in their actions with regard to funding and transfer of technology.

11. The Parties shall, in accordance with Article 10, take into consideration in the implementation of the commitments of the Convention the situation of Parties, particularly developing country Parties, with economies that are vulnerable to the adverse effects of the implementation of measures to respond to climate change. This applies notably to Parties with economies that are highly dependent on income generated from the production, processing and export, and/or consumption of fossil fuels and associated energy-intensive products and/or the use of fossil fuels for which such Parties have serious difficulties in switching to alternatives.

Article 5: Research and Systematic Observation

In carrying out their commitments under Article 4, paragraph 1(g), the Parties shall:

(a) Support and further develop, as appropriate, international and intergovernmental programmes and networks or organizations aimed at defining, conducting, assessing and financing research, data collection and systematic observation, taking into account the need to minimize duplication of effort;

(b) Support international and intergovernmental efforts to strengthen systematic observation and national scientific and technical research capacities and capabilities, particularly in developing countries, and to promote access to, and the exchange of, data and analyses thereof obtained from areas beyond national jurisdiction; and

(c) Take into account the particular concerns and needs of developing countries and cooperate in improving their endogenous capacities and capabilities to participate in the efforts referred to in subparagraphs (a) and (b) above.

Article 6: Education, Training and Public Awareness

In carrying out their commitments under Article 4, paragraph 1(i), the Parties shall:

(a) Promote and facilitate at the national and, as appropriate, subregional and regional levels, and in accordance with national laws and regulations, and within their respective capacities:

(i) The development and implementation of educational and public awareness programmes on climate change and its effects;

(ii) Public access to information on climate change and its effects;

(iii) Public participation in addressing climate change and its effects and developing adequate responses; and

(iv) Training of scientific, technical and managerial personnel.

(b) Cooperate in and promote, at the international level, and, where appropriate, using existing bodies:

(i) The development and exchange of educational and public awareness material on climate change and its effects; and

(ii) The development and implementation of education and training programmes, including the strengthening of national institutions and the exchange or secondment of personnel to train experts in this field, in particular for developing countries.

Article 7: Conference of the Parties

1. A Conference of the Parties is hereby established.

2. The Conference of the Parties, as the supreme body of this Convention, shall keep under regular review the implementation of the Convention and any related legal instruments that the Conference of the Parties may adopt, and shall make, within its mandate, the decisions necessary to promote the effective implementation of the Convention. To this end, it shall:

   (a) Periodically examine the obligations of the Parties and the institutional arrangements under the Convention, in the light of the objective of the Convention, the experience gained in its implementation and the evolution of scientific and technological knowledge;

   (b) Promote and facilitate the exchange of information on measures adopted by the Parties to address climate change and its effects, taking into account the differing circumstances, responsibilities and capabilities of the Parties and their respective commitments under the Convention;

   (c) Facilitate, at the request of two or more Parties, the coordination of measures adopted by them to address climate change and its effects, taking into account the differing circumstances, responsibilities and capabilities of the Parties and their respective commitments under the Convention;

   (d) Promote and guide, in accordance with the objective and provisions of the Convention, the development and periodic refinement of comparable methodologies, to be agreed on by the Conference of the Parties, inter alia, for preparing inventories of greenhouse gas emissions by sources and removals by sinks, and for evaluating the effectiveness of measures to limit the emissions and enhance the removals of these gases;

(e) Assess, on the basis of all information made available to it in accordance with the provisions of the Convention, the implementation of the Convention by the Parties, the overall effects of the measures taken pursuant to the Convention, in particular environmental, economic and social effects as well as their cumulative impacts and the extent to which progress towards the objective of the Convention is being achieved;

(f) Consider and adopt regular reports on the implementation of the Convention and ensure their publication;

(g) Make recommendations on any matters necessary for the implementation of the Convention;

(h) Seek to mobilize financial resources in accordance with Article 4, paragraphs 3, 4 and 5, and Article 11;

(i) Establish such subsidiary bodies as are deemed necessary for the implementation of the Convention;

(j) Review reports submitted by its subsidiary bodies and provide guidance to them;

(k) Agree upon and adopt, by consensus, rules of procedure and financial rules for itself and for any subsidiary bodies;

(l) Seek and utilize, where appropriate, the services and cooperation of, and information provided by, competent international organizations and intergovernmental and non-governmental bodies; and

(m) Exercise such other functions as are required for the achievement of the objective of the Convention as well as all other functions assigned to it under the Convention.

3. The Conference of the Parties shall, at its first session, adopt its own rules of procedure as well as those of the subsidiary bodies established by the Convention, which shall include decision-making procedures for matters not already covered by decision- making procedures stipulated in the Convention. Such procedures may include specified majorities required for the adoption of particular decisions.

4. The first session of the Conference of the Parties shall be convened by the interim secretariat referred to in Article 21 and shall take place not later than one year after the date of entry into force of the Convention. Thereafter, ordinary sessions of the Conference of the Parties shall be held every year unless otherwise decided by the Conference of the Parties.

5. Extraordinary sessions of the Conference of the Parties shall be held at such other times as may be deemed necessary by the Conference, or at the written request of any Party, provided that, within six months of the request being communicated to the Parties by the secretariat, it is supported by at least one third of the Parties.

6. The United Nations, its specialized agencies and the International Atomic Energy Agency, as well as any State member thereof or observers thereto not Party to the

Convention, may be represented at sessions of the Conference of the Parties as observers. Any body or agency, whether national or international, governmental or non-governmental, which is qualified in matters covered by the Convention, and which has informed the secretariat of its wish to be represented at a session of the Conference of the Parties as an observer, may be so admitted unless at least one third of the Parties present object. The admission and participation of observers shall be subject to the rules of procedure adopted by the Conference of the Parties.

Article 8: Secretariat

1. A secretariat is hereby established.

2. The functions of the secretariat shall be:

   (a) To make arrangements for sessions of the Conference of the Parties and its subsidiary bodies established under the Convention and to provide them with services as required;

   (b) To compile and transmit reports submitted to it;

   (c) To facilitate assistance to the Parties, particularly developing country Parties, on request, in the compilation and communication of information required in accordance with the provisions of the Convention;

   (d) To prepare reports on its activities and present them to the Conference of the Parties;

   (e) To ensure the necessary coordination with the secretariats of other relevant international bodies;

   (f) To enter, under the overall guidance of the Conference of the Parties, into such administrative and contractual arrangements as may be required for the effective discharge of its functions; and

(g) To perform the other secretariat functions specified in the Convention and in any of its protocols and such other functions as may be determined by the Conference of the Parties.

3. The Conference of the Parties, at its first session, shall designate a permanent secretariat and make arrangements for its functioning.

Article 9: Subsidiary Body for Scientific and Technological Advice

1. A subsidiary body for scientific and technological advice is hereby established to provide the Conference of the Parties and, as appropriate, its other subsidiary bodies with timely information and advice on scientific and technological matters relating to the Convention. This body shall be open to participation by all Parties and shall be multidisciplinary. It shall comprise government representatives competent in the relevant field of expertise. It shall report regularly to the Conference of the Parties on all aspects of its work.

2. Under the guidance of the Conference of the Parties, and drawing upon existing competent international bodies, this body shall:

   (a) Provide assessments of the state of scientific knowledge relating to climate change and its effects;

   (b) Prepare scientific assessments on the effects of measures taken in the implementation of the Convention;

   (c) Identify innovative, efficient and state-of-the-art technologies and know-how and advise on the ways and means of promoting development and/or transferring such technologies;

(d) Provide advice on scientific programmes, international cooperation in research and development related to climate change, as well as on ways and means of supporting endogenous capacity-building in developing countries; and

(e) Respond to scientific, technological and methodological questions that the Conference of the Parties and its subsidiary bodies may put to the body.

3. The functions and terms of reference of this body may be further elaborated by the Conference of the Parties.

Article 10: Subsidiary Body for Implementation

1. A subsidiary body for implementation is hereby established to assist the Conference of the Parties in the assessment and review of the effective implementation of the Convention. This body shall be open to participation by all Parties and comprise government representatives who are experts on matters related to climate change. It shall report regularly to the Conference of the Parties on all aspects of its work.

2. Under the guidance of the Conference of the Parties, this body shall:

(a) Consider the information communicated in accordance with Article 12, paragraph 1, to assess the overall aggregated effect of the steps taken by the Parties in the light of the latest scientific assessments concerning climate change;

(b) Consider the information communicated in accordance with Article 12, paragraph 2, in order to assist the Conference of the Parties in carrying out the reviews required by Article 4, paragraph 2(d); and

(c) Assist the Conference of the Parties, as appropriate, in the preparation and implementation of its decisions.

Article 11: Financial Mechanism

1. A mechanism for the provision of financial resources on a grant or concessional basis, including for the transfer of technology, is hereby defined. It shall function under the guidance of and be accountable to the Conference of the Parties, which shall decide on its policies, programme priorities and eligibility criteria related to this Convention. Its operation shall be entrusted to one or more existing international entities.

2. The financial mechanism shall have an equitable and balanced representation of all Parties within a transparent system of governance.

3. The Conference of the Parties and the entity or entities entrusted with the operation of the financial mechanism shall agree upon arrangements to give effect to the above paragraphs, which shall include the following:

   (a) Modalities to ensure that the funded projects to address climate change are in conformity with the policies, programme priorities and eligibility criteria established by the Conference of the Parties;

   (b) Modalities by which a particular funding decision may be reconsidered in light of these policies, programme priorities and eligibility criteria;

   (c) Provision by the entity or entities of regular reports to the Conference of the Parties on its funding operations, which is consistent with the requirement for accountability set out in paragraph 1 above; and

(d) Determination in a predictable and identifiable manner of the amount of funding necessary and available for the implementation of this Convention and the conditions under which that amount shall be periodically reviewed.

4. The Conference of the Parties shall make arrangements to implement the above- mentioned provisions at its first session, reviewing and taking into account the interim arrangements referred to in Article 21, paragraph 3, and shall decide whether these interim arrangements shall be maintained. Within four years thereafter, the Conference of the Parties shall review the financial mechanism and take appropriate measures.

5. The developed country Parties may also provide and developing country Parties avail themselves of, financial resources related to the implementation of the Convention through bilateral, regional and other multilateral channels.

Article 12: Communication of Information Related to Implementation

1. In accordance with Article 4, paragraph 1, each Party shall communicate to the Conference of the Parties, through the secretariat, the following elements of information:

   (a) A national inventory of anthropogenic emissions by sources and removals by sinks of all greenhouse gases not controlled by the Montreal Protocol, to the extent its capacities permit, using comparable methodologies to be promoted and agreed upon by the Conference of the Parties;

(b) A general description of steps taken or envisaged by the Party to implement the Convention; and

(c) Any other information that the Party considers relevant to the achievement of the objective of the Convention and suitable for inclusion in its communication, including, if feasible, material relevant for calculations of global emission trends.

2. Each developed country Party and each other Party included in Annex I shall incorporate in its communication the following elements of information:

(a) A detailed description of the policies and measures that it has adopted to implement its commitment under Article 4, paragraphs 2(a) and 2(b); and

(b) A specific estimate of the effects that the policies and measures referred to in subparagraph (a) immediately above will have on anthropogenic emissions by its sources and removals by its sinks of greenhouse gases during the period referred to in Article 4, paragraph 2(a).

3. In addition, each developed country Party and each other developed Party included in Annex II shall incorporate details of measures taken in accordance with Article 4, paragraphs 3, 4 and 5.

4. Developing country Parties may, on a voluntary basis, propose projects for financing, including specific technologies, materials, equipment, techniques or practices that would be needed to implement such projects, along with, if possible, an estimate of all incremental costs, of the reductions of emissions and increments of removals of greenhouse gases, as well as an estimate of the consequent benefits.

5. Each developed country Party and each other Party included in Annex I shall make its initial communication within six months of the entry into force of the Convention for that Party. Each Party not so listed shall make its initial communication within three years of the entry into force of the Convention for that Party, or of the availability of financial resources in accordance with Article 4, paragraph 3. Parties that are least developed countries may make their initial communication at their discretion. The frequency of subsequent communications by all Parties shall be determined by the Conference of the Parties, taking into account the differentiated timetable set by this paragraph.

6. Information communicated by Parties under this Article shall be transmitted by the secretariat as soon as possible to the Conference of the Parties and to any subsidiary bodies concerned. If necessary, the procedures for the communication of information may be further considered by the Conference of the Parties.

7. From its first session, the Conference of the Parties shall arrange for the provision to developing country Parties of technical and financial support, on request, in compiling and communicating information under this Article, as well as in identifying the technical and financial needs associated with proposed projects and response measures under Article 4. Such support may be provided by other Parties, by competent international organizations and by the secretariat, as appropriate.

8. Any group of Parties may, subject to guidelines adopted by the Conference of the Parties, and to prior notification to the Conference of the Parties, make a joint communication in fulfilment of their obligations under this Article, provided that such a communication includes

information on the fulfilment by each of these Parties of its individual obligations under the Convention.

9. Information received by the secretariat that is designated by a Party as confidential, in accordance with criteria to be established by the Conference of the Parties, shall be aggregated by the secretariat to protect its confidentiality before being made available to any of the bodies involved in the communication and review of information.

10. Subject to paragraph 9 above, and without prejudice to the ability of any Party to make public its communication at any time, the secretariat shall make communications by Parties under this Article publicly available at the time they are submitted to the Conference of the Parties.

Article 13: Resolution of Questions Regarding Implementation

The Conference of the Parties shall, at its first session, consider the establishment of a multilateral consultative process, available to Parties on their request, for the resolution of questions regarding the implementation of the Convention.

Article 14: Settlement of Disputes

1. In the event of a dispute between any two or more Parties concerning the interpretation or application of the Convention, the Parties concerned shall seek a settlement of the dispute through negotiation or any other peaceful means of their own choice.

2. When ratifying, accepting, approving or acceding to the Convention, or at any time thereafter, a Party which is not a regional economic integration organization may declare in a written instrument submitted to the Depositary that, in respect of any dispute concerning the interpretation or application of the Convention, it

recognizes as compulsory ipso facto and without special agreement, in relation to any Party accepting the same obligation:

(a) Submission of the dispute to the International Court of Justice, and/or

(b) Arbitration in accordance with procedures to be adopted by the Conference of the Parties as soon as practicable, in an annex on arbitration.

A Party which is a regional economic integration organization may make a declaration with like effect in relation to arbitration in accordance with the procedures referred to in subparagraph (b) above.

3. A declaration made under paragraph 2 above shall remain in force until it expires in accordance with its terms or until three months after written notice of its revocation has been deposited with the Depositary.

4. A new declaration, a notice of revocation or the expiry of a declaration shall not in any way affect proceedings pending before the International Court of Justice or the arbitral tribunal, unless the parties to the dispute otherwise agree.

5. Subject to the operation of paragraph 2 above, if after twelve months following notification by one Party to another that a dispute exists between them, the Parties concerned have not been able to settle their dispute through the means mentioned in paragraph 1 above, the dispute shall be submitted, at the request of any of the parties to the dispute, to conciliation.

6. A conciliation commission shall be created upon the request of one of the parties to the dispute. The

commission shall be composed of an equal number of members appointed by each party concerned and a chairman chosen jointly by the members appointed by each party. The commission shall render a recommendatory award, which the parties shall consider in good faith.

7. Additional procedures relating to conciliation shall be adopted by the Conference of the Parties, as soon as practicable, in an annex on conciliation.

8. The provisions of this Article shall apply to any related legal instrument which the Conference of the Parties may adopt, unless the instrument provides otherwise.

Article 15: Amendments to the Convention

1. Any Party may propose amendments to the Convention.

2. Amendments to the Convention shall be adopted at an ordinary session of the Conference of the Parties. The text of any proposed amendment to the Convention shall be communicated to the Parties by the secretariat at least six months before the meeting at which it is proposed for adoption. The secretariat shall also communicate proposed amendments to the signatories to the Convention and, for information, to the Depositary.

3. The Parties shall make every effort to reach agreement on any proposed amendment to the Convention by consensus. If all efforts at consensus have been exhausted, and no agreement reached, the amendment shall as a last resort be adopted by a three-fourths majority vote of the Parties present and voting at the meeting. The adopted amendment shall be communicated by the secretariat to the Depositary, who shall circulate it to all Parties for their acceptance.

4. Instruments of acceptance in respect of an amendment shall be deposited with the Depositary. An amendment adopted in accordance with paragraph 3 above shall enter into force for those Parties having accepted it on the ninetieth day after the date of receipt by the Depositary of an instrument of acceptance by at least three fourths of the Parties to the Convention.

5. The amendment shall enter into force for any other Party on the ninetieth day after the date on which that Party deposits with the Depositary its instrument of acceptance of the said amendment.

6. For the purposes of this Article, "Parties present and voting" means Parties present and casting an affirmative or negative vote.

Article 16 Adoption and Amendment of Annexes to the Convention

1. Annexes to the Convention shall form an integral part thereof and, unless otherwise expressly provided, a reference to the Convention constitutes at the same time a reference to any annexes thereto. Without prejudice to the provisions of Article 14, paragraphs 2(b) and 7, such annexes shall be restricted to lists, forms and any other material of a descriptive nature that is of a scientific, technical, procedural or administrative character.

2. Annexes to the Convention shall be proposed and adopted in accordance with the procedure set forth in Article 15, paragraphs 2, 3 and 4.

3. An annex that has been adopted in accordance with paragraph 2 above shall enter into force for all Parties to the Convention six months after the date of the communication by the Depositary to such Parties of the

adoption of the annex, except for those Parties that have notified the Depositary, in writing, within that period of their non-acceptance of the annex. The annex shall enter into force for Parties which withdraw their notification of non-acceptance on the ninetieth day after the date on which withdrawal of such notification has been received by the Depositary.

4. The proposal, adoption and entry into force of amendments to annexes to the Convention shall be subject to the same procedure as that for the proposal, adoption and entry into force of annexes to the Convention in accordance with paragraphs 2 and 3 above.

5. If the adoption of an annex or an amendment to an annex involves an amendment to the Convention, that annex or amendment to an annex shall not enter into force until such time as the amendment to the Convention enters into force.

Article 17: Protocols

1. The Conference of the Parties may, at any ordinary session, adopt protocols to the Convention.

2. The text of any proposed protocol shall be communicated to the Parties by the secretariat at least six months before such a session.

3. The requirements for the entry into force of any protocol shall be established by that instrument.

4. Only Parties to the Convention may be Parties to a protocol.

5. Decisions under any protocol shall be taken only by the Parties to the protocol concerned.

Article 18: Right to Vote

1. Each Party to the Convention shall have one vote, except as provided for in paragraph 2 below.

2. Regional economic integration organizations, in matters within their competence, shall exercise their right to vote with a number of votes equal to the number of their member States that are Parties to the Convention. Such an organization shall not exercise its right to vote if any of its member States exercises its right, and vice versa.

Article 19: Depository

The Secretary-General of the United Nations shall be the Depositary of the Convention and of protocols adopted in accordance with Article 17.

Article 20: Signiture

This Convention shall be open for signature by States Members of the United Nations or of any of its specialized agencies or that are Parties to the Statute of the International Court of Justice and by regional economic integration organizations at Rio de Janeiro, during the United Nations Conference on Environment and Development, and thereafter at United Nations Headquarters in New York from 20 June 1992 to 19 June 1993.

Article 21: Interim Arrangements

1. The secretariat functions referred to in Article 8 will be carried out on an interim basis by the secretariat established by the General Assembly of the United Nations in its resolution 45/212 of 21 December 1990, until the completion of the first session of the Conference of the Parties.

2. The head of the interim secretariat referred to in paragraph 1 above will cooperate closely with the

Intergovernmental Panel on Climate Change to ensure that the Panel can respond to the need for objective scientific and technical advice. Other relevant scientific bodies could also be consulted.

3. The Global Environment Facility of the United Nations Development Programme, the United Nations Environment Programme and the International Bank for Reconstruction and Development shall be the international entity entrusted with the operation of the financial mechanism referred to in Article 11 on an interim basis. In this connection, the Global Environment Facility should be appropriately restructured and its membership made universal to enable it to fulfil the requirements of Article 11.

Article 22: Ratification, Acceptance, Approval or Accession

1. The Convention shall be subject to ratification, acceptance, approval or accession by States and by regional economic integration organizations. It shall be open for accession from the day after the date on which the Convention is closed for signature. Instruments of ratification, acceptance, approval or accession shall be deposited with the Depositary.

2. Any regional economic integration organization which becomes a Party to the Convention without any of its member States being a Party shall be bound by all the obligations under the Convention. In the case of such organizations, one or more of whose member States is a Party to the Convention, the organization and its member States shall decide on their respective responsibilities for the performance of their obligations under the Convention. In such cases, the organization and the member States shall not be entitled to exercise rights under the Convention concurrently.

3. In their instruments of ratification, acceptance, approval or accession, regional economic integration organizations shall declare the extent of their competence with respect to the matters governed by the Convention. These organizations shall also inform the Depositary, who shall in turn inform the Parties, of any substantial modification in the extent of their competence.

Article 23: Entry into Force

1. The Convention shall enter into force on the ninetieth day after the date of deposit of the fiftieth instrument of ratification, acceptance, approval or accession.

2. For each State or regional economic integration organization that ratifies, accepts or approves the Convention or accedes thereto after the deposit of the fiftieth instrument of ratification, acceptance, approval or accession, the Convention shall enter into force on the ninetieth day after the date of deposit by such State or regional economic integration organization of its instrument of ratification, acceptance, approval or accession.

3. For the purposes of paragraphs 1 and 2 above, any instrument deposited by a regional economic integration organization shall not be counted as additional to those deposited by States members of the organization.

Article 24: Reservations

No reservations may be made to the Convention.

Article 25: Withdrawal

1. At any time after three years from the date on which the Convention has entered into force for a Party, that Party may withdraw from the Convention by giving written notification to the Depositary.

2. Any such withdrawal shall take effect upon expiry of one year from the date of receipt by the Depositary of the notification of withdrawal, or on such later date as may be specified in the notification of withdrawal.

3. Any Party that withdraws from the Convention shall be considered as also having withdrawn from any protocol to which it is a Party.

Index

E

J

K